Methoden zur Anpassung virtueller Arbeitsumgebungen an die Herausforderungen effizienter Digitalisierungsstrategien in der synthetischen Chemie

Methods for adapting virtual environments to the challenges of efficient digitization strategies in synthetic chemistry

Zur Erlangung des akademischen Grades eines

DOKTORS DER NATURWISSENSCHAFTEN

(Dr. rer. nat.)

bei der Fakultät für Chemie und Biowissenschaften
des Karlsruher Institut für Technologie (KIT)
genehmigte

DISSERTATION

von

Yu-Chieh Huang
aus Taipeh, Taiwan

Dekan: Prof. Dr. Reinhard Fischer
Referent: Prof. Dr. Stefan Bräse
Korreferent: Prof. Dr. Ralf H. Reussner
Tag der mündlichen Prüfung: 17.10.2019

Band 83
Beiträge zur organischen Synthese
Hrsg.: Stefan Bräse

Prof. Dr. Stefan Bräse
Institut für Organische Chemie
Karlsruher Institut für Technologie (KIT)
Fritz-Haber-Weg 6
D-76131 Karlsruhe

Bibliographic information published by the Deutsche Nationalbibliothek

The Deutsche Nationalbibliothek lists this publication in the Deutsche Nationalbibliografie; detailed bibliographic data are available in the Internet at http://dnb.d-nb.de

ISBN 978-3-8325-5052-3
ISSN 1862-5681

Logos Verlag Berlin GmbH
Comeniushof, Gubener Str. 47,
10243 Berlin
Tel.: +49 030 42 85 10 90
Fax: +49 030 42 85 10 92
INTERNET: http://www.logos-verlag.de

Für meine Familie,
meine Freunde
und mich.

Phantasie ist wichtiger als Wissen, denn Wissen ist begrenzt.

\- Albert Einstein

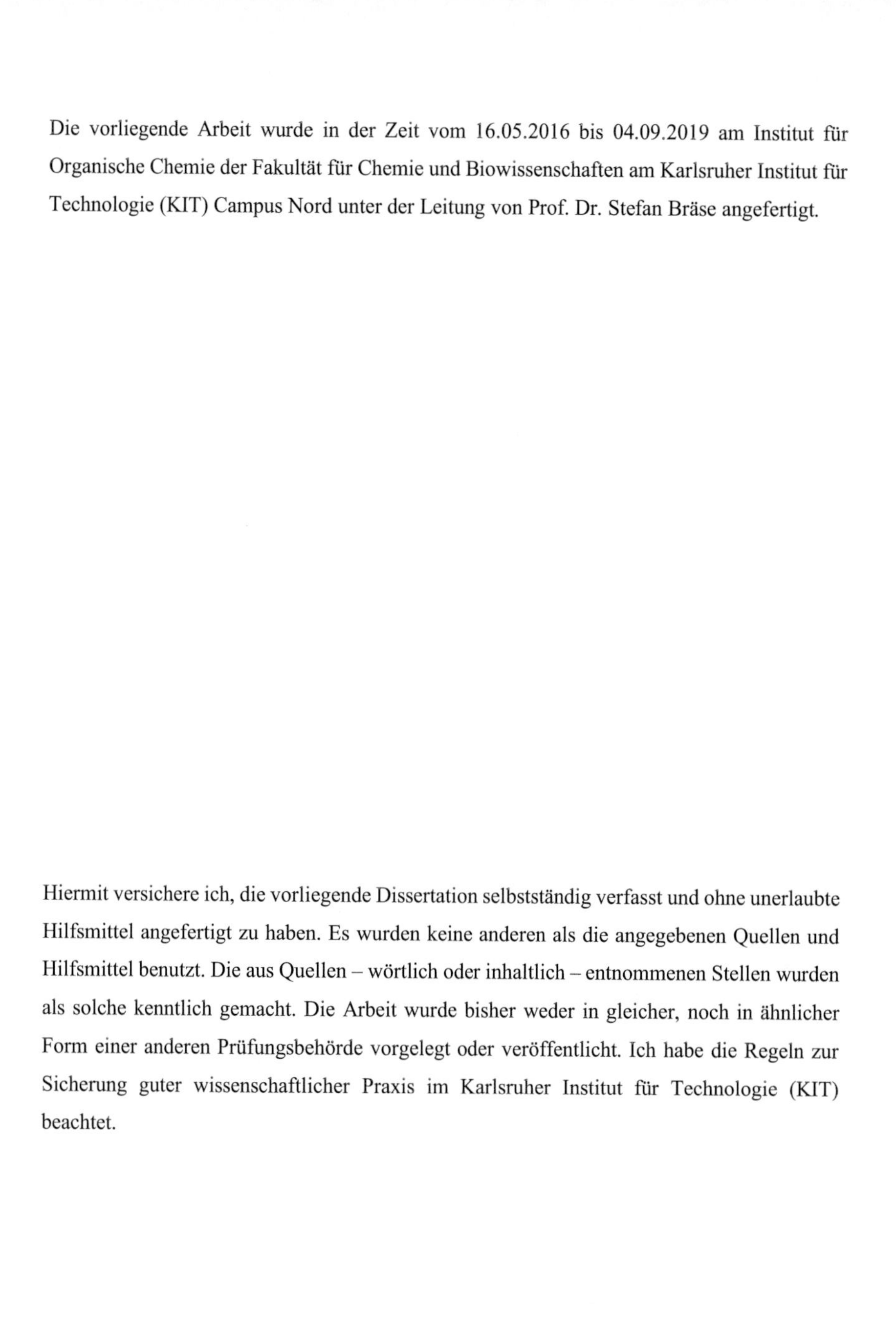

Die vorliegende Arbeit wurde in der Zeit vom 16.05.2016 bis 04.09.2019 am Institut für Organische Chemie der Fakultät für Chemie und Biowissenschaften am Karlsruher Institut für Technologie (KIT) Campus Nord unter der Leitung von Prof. Dr. Stefan Bräse angefertigt.

Hiermit versichere ich, die vorliegende Dissertation selbstständig verfasst und ohne unerlaubte Hilfsmittel angefertigt zu haben. Es wurden keine anderen als die angegebenen Quellen und Hilfsmittel benutzt. Die aus Quellen – wörtlich oder inhaltlich – entnommenen Stellen wurden als solche kenntlich gemacht. Die Arbeit wurde bisher weder in gleicher, noch in ähnlicher Form einer anderen Prüfungsbehörde vorgelegt oder veröffentlicht. Ich habe die Regeln zur Sicherung guter wissenschaftlicher Praxis im Karlsruher Institut für Technologie (KIT) beachtet.

Summary

In the context of this dissertation, methods were investigated that can contribute to a successful digitization strategy in experimental synthetic chemistry. Three areas were identified which could support a change in the current documentation and working methods. These are (1) the development of software for the processing of spectroscopic data and the comparison of the extracted results with the in silico predictions of the target compounds, (2) the generation of ML-based predictions for reaction control (reaction templates and reaction temperature) and (3) the automatic generation of reports from entries of an electronic laboratory journal. The developed models will be discussed and analyzed.

ChemSpectra is a service that enables the processing and analysis of infrared (IR), nuclear magnetic resonance (NMR) and mass spectra. Deep learning models are trained to infer the existence of functional groups from the IR profile. Currently, 36 of the 44 most common functional groups are trained with balanced accuracies of at least 80%. Regarding NMR data, the NMRShiftDB service is used to semi-automatically evaluate signals from NMR spectra. The automatic evaluation of the MS spectra is done by identifying the corresponding mass-charge ratio in the spectrum.

Furthermore, a method was developed to evaluate the quality of the classification of reactions based on their reaction templates. Such a classification by extraction of reaction templates is especially important for cheminformatics projects. High-quality extraction methods can be used to make temperature and reaction predictions, for example. The application of the obtained methods resulted in an accuracy of the temperature prediction of about 73%. The current accuracy of the reaction template prediction is about 77%.

The last model allows the automatic generation of reports from predefined entries of an electronic laboratory journal. The reporting tool developed with ChemDraw integration makes it possible to structure documents in such a way that chemists do not have to change their traditional documentation procedures including the use of ChemDraw. The created tool allows the modification of the received reports including the chemical structures even after the completion in Word. In addition to a standard template for internal reports, other formats for implementing the FAIR data principles are supported.

Zusammenfassung

Im Rahmen dieser Dissertation wurden Methoden untersucht, die zu einer erfolgreichen Digitalisierungsstrategie in der experimentellen synthetischen Chemie beitragen können. Es wurden drei Bereiche identifiziert, die hier insbesondere eine Veränderung der aktuellen Dokumentations- und Arbeitsweise unterstützen können. Dies sind (1) die Entwicklung von Software zur Bearbeitung von spektroskopischen Daten und dem Ableich der extrahierten Ergebnisse mit den in silico-Vorhersagen der Zielverbindungen, (2) die Generierung von ML-basierten Vorhersagen für die Reaktionsführung (Reaktionstemplates und Reaktionstemperatur) und (3) die automatische Erstellung von Berichten aus Einträgen eines elektronischen Laborjournals. Die entwickelten Modelle werden jeweils diskutiert und analysiert.

ChemSpectra ist ein Dienst, der die Bearbeitung und Analyse von Infrarot- (IR), Kernresonanz- (NMR) und Massen-Spektren ermöglicht. Deep Learning Modelle werden trainiert, um aus dem IR-Profil auf die Existenz von funktionellen Gruppen zu schließen. Derzeit werden 36 der 44 häufigsten funktionellen Gruppen mit ausgewogenen Genauigkeiten von mindestens 80% trainiert. Hinsichtlich der NMR-Daten wird der Service der NMRShiftDB genutzt, um die Signale aus NMR-Spektren halbautomatisch auszuwerten. Die automatische Auswertung der MS-Spektren erfolgt durch Identifizierung des entsprechenden Masse-Ladungs-Verhältnisses im Spektrum.

Weiterhin wurde eine Methode entwickelt, um die Qualität der Klassifizierung von Reaktionen anhand ihrer Reaktionstemplates zu bewerten. Eine solche Klassifizierung durch Extraktion von Reaktionstemplates ist insbesondere für chemoinformatische Vorhaben wichtig. Durch qualitativ hochwertige Extraktionsverfahren können z.B. Temperatur- und Reaktionsvorhersagen ermöglicht werden. Die Anwendung der erhaltenen Verfahren ergab eine Genauigkeit der Temperaturvorhersage von etwa 73%. Die aktuelle Genauigkeit der Reaktionstemplate-Vorhersage liegt bei etwa 77%.

Das letzte Modell ermöglicht die automatische Erstellung von Berichten aus vorgegebenen Einträgen eines elektronischen Laborjournals. Das entwickelte Reporting-Tool mit ChemDraw Einbindung ermöglicht es, Dokumente so aufzubauen, dass Chemiker ihre traditionellen Dokumentationsverfahren inklusive der Nutzung von ChemDraw nicht ändern müssen. Das erstellte Tool erlaubt die Änderung der erhaltenen Reports inklusive der chemischen Strukturen auch nach der Fertigstellung in Word. Neben einem Standard-Template für interne Berichte werden weitere Formate zur Umsetzung der FAIR data Prinzipien unterstützt.

Table of Contents

1 Introduction *1*

2 Background *3*

2.1 Cheminformatics 3

2.2 Structure Elucidation by Spectra 4

2.2.1 Spectrum Viewers and Editors 5

2.2.2 Structure Elucidation Methods 6

2.3 Synthesis and Retrosynthetic Predictions 9

2.3.1 Template Model 10

2.3.2 Translation Model 14

2.3.3 Graph Model 19

2.4 Reporting 22

3 Aim *23*

4 Main Part - ChemSpectra *25*

4.1 Overview 26

4.2 React-spectra-editor 28

4.2.1 Spectrum Page 28

4.2.2 Data Input 32

4.2.3 Control Panel 32

4.2.4 Editing in the Main Editor 33

4.2.5 Analysis Panel 34

4.3 Chem-spectra-client 36

4.4 Chem-spectra-app 39

4.4.1 Spectra Transformation 40

4.4.2 Spectra Inference 45

4.5 Building IR Inference Models 47

4.5.1 Data Source 48

4.5.2 Spectrum Preprocessing 48

4.5.3 Functional Group Extraction 51

4.5.4 Deep Learning Models 52

4.5.5 Model Training and Verification 53

4.6 Chem-spectra-deep-ir 59

4.7 Chemotion-ELN Integration 59

4.8 Quality Assurance – Quality Control and Quick Check 63

4.8.1 Overview 63

4.8.2 Analysis Detail 65

4.9 Discussion 67

5 Main Part - Reaction Prediction *69*

5.1 Reaction Template 70

5.2 Template Extraction 72

5.2.1 Data Cleaning 72

5.2.2 Atom Mapping 73

5.2.3 Core Extraction 74

5.2.4 Neighbor Extraction 75

5.3 Template Verification - Restoration 76

5.4 Predictions 80

5.4.1 Temperature Prediction 80

5.4.2 Template Prediction 83

5.5 Discussion 86

6 Main Part - Reporting *87*

6.1 Report Types 88

6.1.1 Standard Report 89

6.1.2 Supporting Information 93

6.2 User Interface 100

6.3 Libraries 104

6.3.1 RInChI Ruby Binding 104

6.3.2 ChemDraw Embedding 105

6.4 Discussion 111

7 Summary *113*

7.1 ChemSpectra 113

7.2 Reaction Predictions 114

7.3 Reporting 114

8 Abbreviations *117*

9 Literature *121*

10 Appendix *127*

10.1 Accuracies of Functional Group Predictions **127**

10.2 Curriculum Vitae **130**

10.3 Acknowledgements **132**

Table of Figures

Figure 2-1. Common process flow to verify a proposed structure from a measured spectrum. 4
Figure 2-2. Finding NMR shift boundaries of an atom from the database. 8
Figure 2-3. The Morgan fingerprint of phenylalanine, generated by RDKit. Top: phenylalanine; bottom: substructures and corresponding bit indexes are listed. 11
Figure 2-4. A 2-layer MLP architecture. 12
Figure 2-5. Schematic depiction of an RNN architecture. 17
Figure 2-6. The phenylalanine graph. 19
Figure 2-7. The atom adjacency matrix of phenylalanine. 20
Figure 2-8. The bond adjacency matrix of phenylalanine. 20
Figure 4-1. Overview of the ChemSpectra standalone system. 27
Figure 4-2. Overview of the ChemSpectra & the Chemotion-ELN integration. 28
Figure 4-3. An IR spectrum view. 29
Figure 4-4. A ^{1}H NMR spectrum view. 30
Figure 4-5. A ^{13}C NMR spectrum view. 30
Figure 4-6. MS spectrum as a bar graph. 31
Figure 4-7. Data input and process flow. 31
Figure 4-8. Control panel of the React-spectra-editor. 33
Figure 4-9. Analysis panel for an IR spectrum. 35
Figure 4-10. Analysis panel for a ^{13}C NMR spectrum. 36
Figure 4-11. The chem-spectra-client as the proxy to the API server. 37
Figure 4-12. The chem-spectra-clients user interface. 38
Figure 4-13. Flow chart of spectra transformation. 40
Figure 4-14. Flow chart of reading a JCAMP-DX file. 42
Figure 4-15. Mzml and raw conversion to JCAMP-DX inside the python MSConvert class. 43
Figure 4-16. Abstraction of a composed NMR JCAMP-DX file. 45
Figure 4-17. The original spectrum of Chemspider CSID 110120. 50
Figure 4-18. The processed spectrum of Chemspider CSID 110120. 50
Figure 4-19. Phenylalanine functional groups are highlighted by SMARTS matching. 52
Figure 4-20. MLP loss for the 1st frequent functional group, C-,:O. 57
Figure 4-21. MLP balanced accuracy for the 1st frequent functional group, C-,:O. 57
Figure 4-22. MLP loss for the 1st frequent functional group, C-,:O. 58
Figure 4-23. MLP balanced accuracy for the 1st frequent functional group, C-,:O. 58

Figure 4-24. IR inference flow 59
Figure 4-25. Sequence diagram of the state machine 62
Figure 4-26. An example of an overview of the quality assurance tool in the ELN UI 63
Figure 4-27. Analysis details of ^{1}H NMR. 65
Figure 4-28. Analysis details of ^{13}C NMR. 66
Figure 4-29. Analysis details of MS and IR. 66
Figure 5-1. Steps of template extraction 72
Figure 5-2. Atom mapping to identify changes. 74
Figure 5-3. Atom changes with different radius. From top to bottom are radius 0, 1, 2. Pink areas highlight the changes in starting materials, and green parts are those of the product. 76
Figure 5-4. The full template extraction procedure: template extraction with the restoration. 77
Figure 5-5. The distributed computing architecture 79
Figure 5-6. Temperature distribution of a reaction class 81
Figure 5-7. Temperature distribution of a reaction class 81
Figure 5-8. The architecture of temperature prediction 82
Figure 5-9. Validation and testing loss 83
Figure 5-10. Input features for the experimental group. In this case, the highest temperature is room temperature 84
Figure 5-11. Deep learning models adopted in this section. 85
Figure 6-1. A sample section of the standard report. 90
Figure 6-2. Optional information in the standard model can be selected in the setting panel. 91
Figure 6-3. A reaction section in the standard model. 92
Figure 6-4. An example of the primary supporting information. 94
Figure 6-5. A scheme of a general procedure. 95
Figure 6-6. Template replacement of Figure 6-5: ethyl in R1 and R3. R2 is a hydrogen-atom. 96
Figure 6-7. A relation among a general procedure and a synthesis is built by selecting 'role' and drag-and-drop the general procedure to the 'according to' field in the browser. 96
Figure 6-8. An example of the spectra supporting information 98
Figure 6-9. A reaction list in the XLSX format. 99
Figure 6-10. A reaction list in the HTML format. 100
Figure 6-11. Accessing the user interface of the report panel. 100
Figure 6-12. The flow chart for creating a report. 101

Figure 6-13. The “Label” window. 102

Figure 6-14. The “Archive” window 103

Figure 6-15. The bidirectional data flow between ChemDraw and a Docx document. 105

Figure 6-16. The process flow of creating OLE from Molfiles and inserting into Docx files. 110

Figure 6-17. The CDXML to CDX conversion flow. 111

List of Tables

Table 2-1. Functionalities of spectra in structure elucidation....4
Table 2-2. A comparison of spectrum software products....6
Table 2-3. Chemical shifts for ^{13}C NMR.[16,33]....7
Table 2-4. Reaction prediction definition....9
Table 2-5. Synthesis prediction models: deep learning networks and encodings....10
Table 2-6. 10 SMILES of phenylalanine, generated by RDKit....16
Table 2-7. RDKit and Open Babel generate two distinct canonical SMILESs....17
Table 2-8. RNN: The distance from previous states to the current neuron influences the memory effect....18
Table 4-1. Modules of ChemSpectra. FE is frontend, the presentation layer. BE is backend, the data processing layer....26
Table 4-2. IR analysis status vs. condition....35
Table 4-3. NMR analysis status vs. condition....36
Table 4-4. Functions of the chem-spectra-app....39
Table 4-5. Possible input spectrum formats for the chem-spectra-app....39
Table 4-6. Services for spectra analysis....46
Table 4-7. Summary of computer-assisted solutions....47
Table 4-8. Preprocessing methods to standardize spectra....48
Table 4-9. Atom features considered as functional groups.[97]....51
Table 4-10. Explanation of phenylalanine functional group formation....52
Table 4-11. 1st sample splits for training, verification, and testing....54
Table 4-12. 44th sample splits for training, verification, and testing....54
Table 4-13. MLP model architecture....55
Table 4-14. 1D-CNN model architecture. (results: Table 10-1)....56
Table 4-15. State definitions for the attachment table....61
Table 5-1. Two reactions, having the same template, are as assigned to the same class....71
Table 5-2. If an atom belongs to the reaction core, at least one of the following factors is altered during the reaction....75
Table 5-3. Apply the full template extraction process to 2000 reactions from the USPTO database....78
Table 5-4. Performance measurement of the full process time by distributed computing....80
Table 5-5. Accuracies of the control group and experimental groups....85

Table 6-1 Report types 88
Table 6-2. Comparison of the three RInChIKeys. 105
Table 6-3. Alanine in CDX format. 107
Table 6-4. Alanine in CDXML format. 108
Table 6-5. An atom node 110
Table 10-1. Accuracies of functional groups, from 1st ~ 10th. 127
Table 10-2. Accuracies of functional groups, from 11th ~ 28th. 128
Table 10-3. Accuracies of functional groups, from 29th ~ 44th. 129

1 Introduction

The vision of the fourth industrial revolution, Industry 4.0,[1-3] is to further advance the integration of physical operations and virtual simulations. Machines, sensors, and users are connected via network infrastructure, on which information can be transmitted from source to destination seamlessly. After aggregating data, machine learning algorithms could learn key factors from features of existing facts. Data-driven inference systems can predict on new statuses without human assistance. As a result, we can achieve more efficient and adaptable systems.

Digitalization is the basis to approach Industry 4.0 since data integration and transmission via the internet must be in the digitized format.[4] Without numerical representation, inter-linkage between the physical and the virtual worlds breaks. In traditional synthetic chemical research, scientists record their idea, daily work, and results on paper. However, it is common that typos, unidentifiable handwritings, and a lack of measures to verify data distort information. Developing data-driven algorithms and transferring knowledge require laborious and repetitive work to retrieve information from archives. Therefore, electronic lab notebooks (ELN),[5,6] in which input and storage of data are inherently digitized, are built to replace classical paper-based documentation methods.

Consistent, reliable, and comprehensive research data management is a crucial topic as a tremendous amount of digitized data is rapidly created and gathered.[7] Without efficient management, preserved data could be duplicated and scattered in machines. Functional data might also include workflow and environmental parameters which could be easily ignored in the beginning. How to preprocess and collect information which maintains the necessity for future operations is a crucial point when designing and developing management systems.

In this work, an open-sourced and web-based ELN, the Chemotion-ELN,[8-10] is improved with three different modules which are developed and implemented. These modules are (1) chemical spectra editing and verification, (2) reaction template and temperature prediction, and (3) a reporting function. The three of them are deep learning and computer-assisted methods designed to increase users' efficiency in synthetic chemistry research. During implementation, the requirement of an efficient research-data management was taken into consideration. In the end, a virtual work platform for synthetic chemistry research and development was realized.

The work and information in this thesis are applied to the sub-domain of organic chemistry. The extension to other fields, e.g., biochemistry and physical chemistry, is currently under development, since the current settings do not fit to the focus of these sub-domains perfectly.

Before going into further detail, a review of related background is presented in the next chapter.

2 Background

2.1 Cheminformatics

Cheminformatics[11,12] is a discipline merging chemistry and information technology. The very first step of the integration is digitalization. Once we can describe chemistry in digital formats, options for further applications are diverse. These options have an impact on the whole scope of chemistry. Digitalization allows information storage and retrieval, data exploration, the development of virtual laboratories, and many others. Solving problems with computers increases efficiency and decreases repetitive manual work.

Digital information storage and retrieval provide a way that users can get and save data from and to a digital system. This method avoids information loss due to valerian handwritings or unorganized documentation. All information has to be described in machine-readable formats, which is preserved in databases or cloud storage solutions. Furthermore, data can be shared with others easily.

By having digital data as a prerequisite, data mining and exploration can be applied to extract features from massive inputs. Using knowledge-based rules, which are constructed depending on theories, is a way to analyze and filter data. It usually requires seasoned scientists to define them. However, rules will become convoluted in the end because of a large number of conditions and exceptions.

In order to overcome this obstacle, techniques without predefined knowledge like statistical analysis and machine learning can be adopted.[13-15] Learning iterations are purely driven by data since the target is reached by minimizing differences among real and virtual simulated outputs. This technique can foster the explorations of the target space because of the fast computing speed. As a result, laborious rule encoding is no longer required.

In this chapter, several topics in the field of organic chemistry, which could be solved by machines using cheminformatics, are reviewed. Due to the focus of this work, the focus is set on structure elucidation by spectra, reaction prediction, and reporting. Their background, functionality, and difficulty are presented as stepstones to give an overview of the motivation of this thesis. These topics are explored and developed because the fusion between real-world information and virtual environment leads to advantages that are craved by academia and industry.

2.2 Structure Elucidation by Spectra

Experimental spectra readouts are required to verify whether a reaction was successful and the target structure was synthesized during the reaction or not. In this thesis, the spectra discussed include ^{1}H and ^{13}C NMR (Nuclear Magnetic Resonance), IR (Infrared) and MS (mass) spectra of which the function in structure elucidation are listed in Table 2-1.[16]

Table 2-1. Functionalities of spectra in structure elucidation.

Spectrum Type	Functionality
^{1}H NMR	Determine types and numbers of protons in different environments
^{13}C NMR	Determine types and numbers of carbons in different environments
IR	Determine the existence of functional groups
MS	Determine molecular weight and fragments

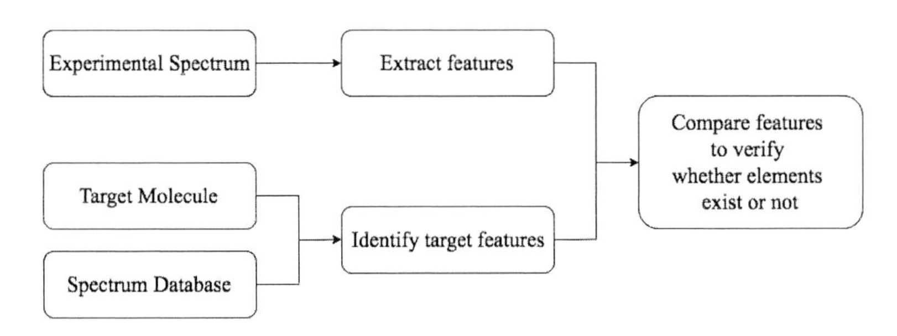

Figure 2-1. Common process flow to verify a proposed structure from a measured spectrum.

Currently, the verification of spectral data is mostly done manually with or without partial computer-assisted methods based on users' experiences.[17-23] Although the spectrum types and mechanisms are not the same, the general workflow is similar. The user needs to visualize the spectra to identify significant features, e.g., features, manually. After the identification, these peaks are compared to other similar molecular structures in knowledge databases, assuming that a similar structure leads to a similar result due to alike physical and chemical compositions. Shown in Figure 2-1 is the usual workflow analyzing a spectrum. Understanding the structure from several types of spectra can provide higher confidence and accuracy.

2.2.1 Spectrum Viewers and Editors

A graphical representation for users to recognized peaks is the very first step of manual feature extraction. Original spectral data may consist of raw data points in frequency or time domain. The software has to have the capability to transform these raw data points into the target domain. Usually, there are around 1000 - 60000 data points, which makes it impossible for humans to read them without rendering scans in an illustrative way. Therefore, a spectrum viewer software is needed.

Editing a spectrum requires additional algorithms or manual control to evaluate data points. Peak picking, assigned manually and/or automatically, locates important features of spectra. The selection by a computer is preferred because the process could require humans to identify the precise position of dozens of points. Meanwhile, adjusting deviation based on the absolute reference is critical. Errors can lead to wrong position values because the whole verification procedure will be affected by deviation. A spectrum editor equipped with an interface for convenient assignment and reference definition is beneficial. Moreover, the calculated features must be updated back to the files for reusable information storage.

Several software packages provide spectra viewing and editing functions: MNova, TopSpin, or Spectragryph are only a few of them. These are commercial software solutions, and there are no or limited free licenses available. Moreover, commercial solutions do not provide a web-based interface that is available to be integrated with other web services. Examples for web-based implementations are NMRPro with SpecdrawJS, JSpecView, jsNMR and SpeckTackle. A comparison of these software products is concluded in Table 2-2.

Table 2-2. A comparison of spectrum software products.

Software	NMR	IR	MS	Open source	Interactive interface	Editing	Integration to other projects
MNova[24,25]	Y		Y		Y	Y	
TopSpin[26]	Y				Y	Y	
Spectragryph[27]		Y			Y	Y	
JSpecView[28]	Y	Y	Y	Y	Y	Limited	
NMRPro/ SpecdrawJS[29]	Y			Y	Y	Y	Y
jsNMR[30]	Y			Y	Y	Limited	Y
SpeckTackle[31,32]	Y	Y	Y	Y	Y		Y

A module with a combined solution for IR, NMR, and MS spectra with complete processing functionalities is missing so far, which currently limits the use of analytical data via web-based systems. NMRPro provides many functionalities, but it is only suitable for NMR spectra. JSpecView can render different types, but its functionalities are limited. A further issue is that it is no longer developed. As a result, the challenge is to build an interface supporting IR, NMR, and MS spectra with reasonable functionalities. A possibility to have continuous development is also essential.

2.2.2 Structure Elucidation Methods

2.2.2.1 NMR

^{1}H and ^{13}C NMR spectra provide information on the chemical environment of proton and carbon atoms, respectively. Two atoms with a similar local environment have similar chemical shifts. Table 2-3 lists molecular substructure versus chemical shifts for ^{13}C NMR as an example for selected functional groups. If alkyl bromides, alkynes, and alkyl fluorides are concerned, they can be identified quickly: shifts are smaller than 60 ppm for alkyl bromides, while alkyl fluorides exhibit shifts greater than 80 ppm. In between, alkynes contribute shifts.

However, the prediction is difficult after adding amines into consideration, since their shifts overlap to alkyl bromides and alkynes. In order to conquer this issue, more conditions need to be added, like relative intensity and shape of the signal. Different local molecular structures will also lead to additional movements or shape changes. This dilemma happens only after taking one more group into account. It will be much more arduous to distinguish them perfectly based on rules when there are many different substructures.

Table 2-3. Chemical shifts for ^{13}C NMR.[16,33]

Functionality	**Chemical shift [ppm]**
alkyl bromides	30 - 60
alkynes	60 - 80
alkyl fluorides	80 - 95
amines	40 - 70

One possible solution is to conclude chemical shifts from a spectrum database and use a data-driven way to solve this. HOSE codes (Hierarchically Ordered Spherical Environment), adopted to identify NMR shifts,[34-36] is a SMILES-like[37,38] descriptor including the target atom and its neighbors. Two atoms with a similar local environment have similar HOSE codes. Figure 2-2 shows a process flow to find shift boundaries of an atom from the database. HOSE codes are extended to deal with stereochemistry with high accuracy.

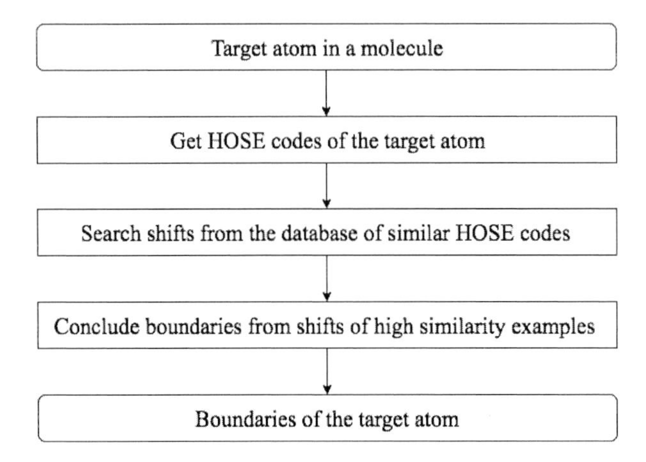

Figure 2-2. Finding NMR shift boundaries of an atom from the database.

This approach, which can reach a decent accuracy, should be integrated into a virtual environment. Such a procedure can lead to an efficient and comprehensive workflow, by which the platform can deliver functions including structure elucidation, data storage, sharing, and reporting. Meanwhile, it is of high interest to use deep learning models to improve this approach.

2.2.2.2 IR

Features in IR spectra indicates to the presence of functional groups. IR spectra were analyzed by machine learning and deep neural network in the past.[39,40] In the given publication, the deep neural network model reaches an accuracy of 79.47%. However, this classification is difficult to scale up and adding new functional groups into consideration requires to retrain the whole model. Moreover, functional groups are defined in a general form and parsed manually. A full data-driven approach is a challenge to achieve by an automatic process.

2.2.2.3 MS

Depending on the machine and method, one MS measurement might generate several scans. The technicians have to inspect them finding the scan with the best quality. This process is done by manual recognition, which is not determined by a numerical value. Therefore, the challenge is to build an algorithm to identify the best scan automatically and numerically.

2.3 Synthesis and Retrosynthetic Predictions

Developing new reactions is a fundamental topic in chemistry, and aims to create new materials breaking existing limitations. There are several factors when planning a reaction: which starting materials should be used and what are possible products (also by-products) of the reaction. Environment conditions, like temperature or humidity, only as an example, may or may not affect the result significantly. Moreover, the yield and throughput of products are also essential measurements to define whether a reaction is beneficial or not.

There are several categories related to this topic: reaction-product prediction, retrosynthetic route planning, and reaction condition prediction. Some definitions are listed in Table 2-4.

Currently, there are three types of reaction prediction models, each of them having different molecular encoding methods to represent molecules owing to the nature of the model. Multiple solutions to tackle the same question arise from how to interpret reactions systematically. Methods to transform reactions and molecules to digital formats are still under active investigation.

Table 2-4. Reaction prediction definition.

Category	Definition
Synthesis product prediction[41-50]	Given starting materials, the system predicts whether there are new products and what the new products are.
Retrosynthesis route planning[45,51-56]	Given products, the system predicts what starting materials are needed to generate them.
Reaction condition prediction[57]	Given starting materials, the system predicts reaction conditions, e.g., temperature.

These essential prediction models are the template model, translation model, and graph model. Table 2-5 lists the models and encodings of synthesis. It has to be mentioned that the solutions to product and condition prediction are slightly different, but the input is the same. For a lucid explanation, phenylalanine, an amino acid with a ring and a stereo bond, is given as a molecule example to explain the encoding in this chapter.

Table 2-5. Synthesis prediction models: deep learning networks and encodings

Model	Deep learning network	Input encoding	Output encoding
Template model	Multi-layer perceptron	Fingerprint	Template class
Translation model	Long-short-term memory	SMILES	SMILES
Graph model	Graph convolutional network	Adjacency matrix & element features	Edit of each atom or bond

2.3.1 Template Model

2.3.1.1 Concept

Several publications[41,45,50] adopt the template model. It is a common scenario to find several reactions having the same element-wise change, which can be considered as the same reaction pattern. This pattern is the reaction template. By knowing starting materials and the template, it is possible to infer products, which is the central concept of this model. The reaction template will be explained more detailed in chapter 5. For this model, the input are molecular fingerprints, while the output are reaction template classes.

2.3.1.2 Fingerprint

The idea of the fingerprint[11,58,59] is to give information on whether molecular fragments exist in the target molecule or not. In order to study the correlation between molecular fragments and reaction tendencies, the fingerprint is one of the most popular features because of its simplicity.

The fingerprint is a bit vector of the molecule. Each bit defines the presence of a substructure, and the value is either 0 or 1 where 1 means the substructure exists in the molecule and 0 means no presence of that substructure. The radius defines the substructure coverage. Atoms outside the radius will not be considered. Figure 2-3 lists the substructures of phenylalanine with both images and corresponding bit indexes, which are generated by RDKit.[60] They are defined using the Morgan fingerprint with a radius of 2. It is the similar to the ECFP4 fingerprint where 4 is the diameter, i.e. radius is 2,.[11,61]

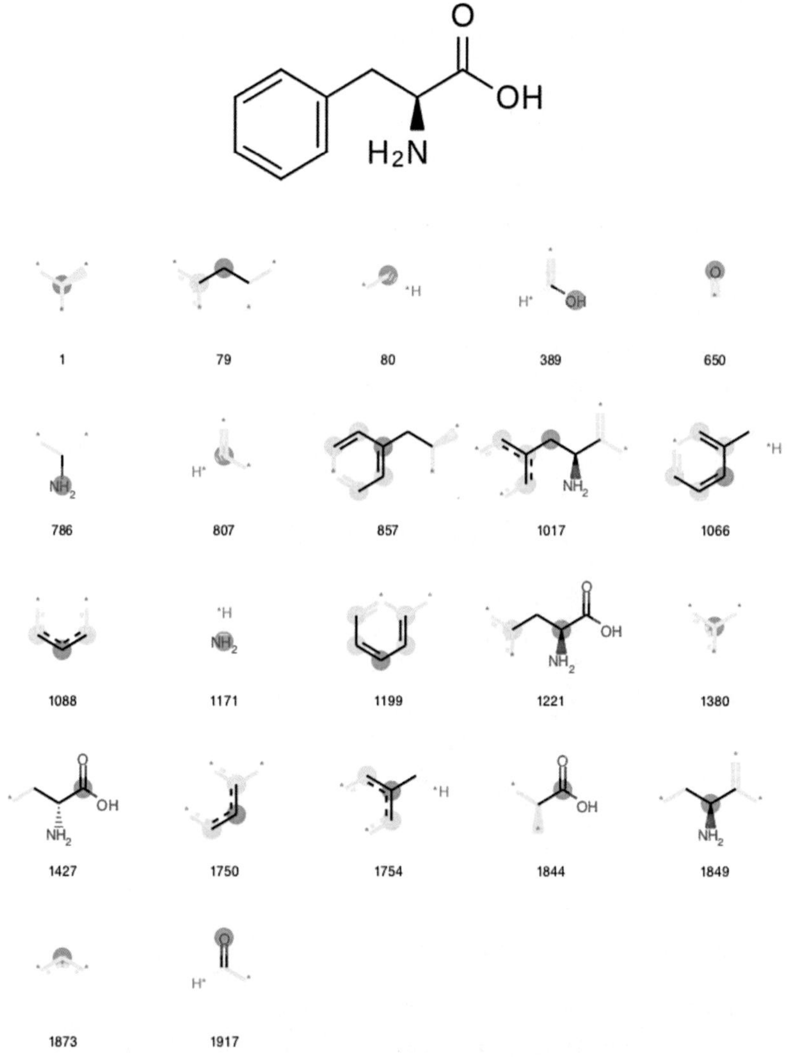

Figure 2-3. The Morgan fingerprint of phenylalanine, generated by RDKit. Top: phenylalanine; bottom: substructures and corresponding bit indexes are listed.

2.3.1.3 MLP (Multi-Layer Perceptron)

A multi-layer perceptron[62] consists of several layers of neural networks. The most straightforward supervised-learning neural network contains an input layer, a hidden layer, and an output layer. For a training set, features are fed into the input layer. The hidden layer contains trainable variables, while the predicted outcome is the linear combination of the input layer and trainable variables passing through a nonlinear active layer. Finally, by comparing the difference between predicted and known results, it is calculated how much the network deviates from the ideal value and trainable variables will be adjusted to decrease the difference.

By stacking several layers of the neural network, a function approximator with high-order nonlinear properties that can deal with more complex problems is achieved. This stacked neural network consists of many layers, and is expanded in the depth direction. Therefore, it is often named as DNN (Deep Neural Network), MLP (Multi-Layer Perceptron) or FFNN (Feed-Forward Neural Network). Figure 2-4 illustrates a 2-layer MLP.

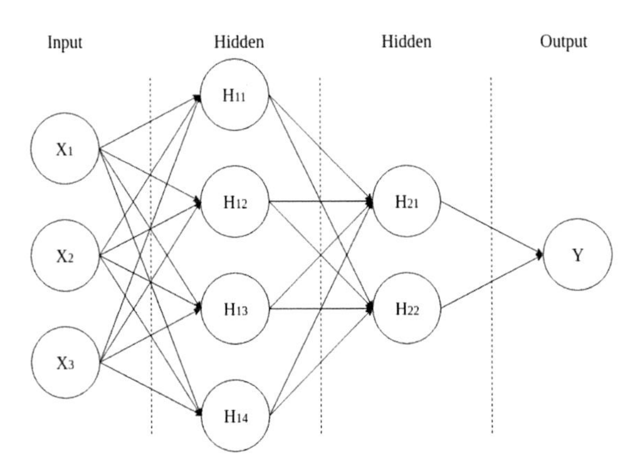

Figure 2-4. A 2-layer MLP architecture.

2.3.1.4 Challenge

The template model predicts template classes. Nevertheless, future inference is limited by available training template classes in the source dataset. Templates outside the training dataset are unable to be predicted by this model since the model is not aware of these. When starting materials of a reaction class, which is absent in the dataset, are inputted, the model will generate many fruitless possibilities which cannot lead to the correct result. Therefore, this model can be used to understand mechanisms similar to published results, but it is not suitable to explore unknown chemical space where reaction templates have not been discovered yet.

Not every reaction template has the same frequency in the database. In a recent work,[45,50] a template was extracted and included in training when there are more than 100 reaction samples for this template. This leads to 8720 templates from the Reaxys database.[63] There are 137 templates which have more than 5000 reaction samples, while around 8650 templates occur in 50 - 100 reactions. This evaluation denotes that many templates are not included in the training.

Imbalanced datasets result in bad training quality of a multi-class classification problem. To illustrate the negative impact of an imbalanced dataset, an extreme scenario will be explained in the following: 100 positives and 4900 negative samples. When the accuracy is the only evaluation parameter, predicting all results as negative will get an accuracy of 98%. The accuracy value is acceptable, but the model does not make any useful prediction.

Using the Reaxys database in publications leads to difficulties in the comparison among different models because it is not publicly available. The analysis using the Reaxys database with the template model considering templates occur more than 100 times has 78% accuracy for the synthesis prediction problem. Other authors reach 33.3% accuracy using the template model on the filtered USPTO dataset with 15000 reactions.[64-66] However, they can reach 68.5% accuracy using a graph model.[41]

Based on the numbers, it seems that the graph model outperforms the template model, but some chemists argue that the template model could achieve higher performance with a larger dataset, e.g., the Reaxys database. The filtered USPTO patent dataset may not contain the best examples because the patent convention is to protect interest but not to publish research data. There is no way to know this unless a public dataset containing all research reactions is available.

Another challenge is the template extraction algorithm. There is no discussion to verify whether the algorithm is accurate or not, and authors only build those template extraction algorithms based on chemical knowledge. Systematically finding pros and cons is not reported yet. The extraction algorithm also depends on mapping tools that have their advantages and disadvantages since no mapping tool alone can tackle all reactions.[67] Moreover, the analysis[45] does not consider the stereochemistry.

2.3.2 Translation Model

2.3.2.1 Concept

The translation model takes advantage of SMILES (simplified molecular-input line-entry system),[49] line notation, which is derived from the sequence ordering of molecule symbols. Its form is similar to a word in human language: each symbol in SMILES represents an atom or a bond, while a word expresses the meaning by prefix, root, and suffix. SMILES is adopted as input and output molecular encodings for translation models in publications,[49,51] in which authors consider that the transformation between starting materials and products in a reaction is similar to translation conversion among two languages. The deep learning model adopted for the translation model is the RNN (Recurrent Neural Network).[68] There are several variations of an RNN, e.g., LSTM (Long-Short-Term Memory).[69,70]

In the translation model, template extraction is not required, which avoids questions owing from mapping and extraction algorithms. Since the translation model can be trained as switching languages between input and output, synthesis and retrosynthesis training are similar by exchanging starting materials and products in input and output.

2.3.2.2 SMILES

By starting from one atom of the molecule, atom symbols are recorded to SMILES one by one until walking through all atoms.[11] Lower-case symbols identify the aromatic ring structure. Single bonds are omitted, and double bonds and triple bonds are (=) and (#), respectively. Moreover, to represent a tetrahedral carbon as a stereocenter, (@) and (@@) denote counter-clockwise and clockwise configurations.

There is one problem when encoding a 2D or 3D structure into a 1D line string: SMILES will be changed when picking a different starting atom. A different branching scheme will also lead to divergence. Ten SMILES of phenylalanine are listed in Table 2-6, which demonstrates a multiple-SMILES-to-one-molecule relationship.

The uniqueness issue of SMILES can be solved by introducing the ordering algorithm, i.e., picking the starting point and selecting branches with predefined rules. This unique SMILES is the canonical SMILES. However, there is no standard ordering algorithm. Two programs can generate two different canonical SMILESs. Nevertheless, as long as the same software does the processing, this deficiency can be avoided. An example of the canonical SMILESs of phenylalanine generated from RDKit[60] and Open Babel[71] is described in

Table 2-7. Two canonical SMILESs are not the same.

Table 2-6. 10 SMILES of phenylalanine, generated by RDKit.

molecule	**SMILES**
phenylalanine	N[C@@H](Cc1ccccc1)C(=O)O
	C(=O)(O)[C@@H](N)Cc1ccccc1
	C([C@@H](C(=O)O)N)c1ccccc1
	C(c1ccccc1)[C@@H](C(=O)O)N
	O=C(O)[C@@H](N)Cc1ccccc1
	[C@@H](C(=O)O)(Cc1ccccc1)N
	[C@H](N)(Cc1ccccc1)C(O)=O
	[C@H](Cc1ccccc1)(C(=O)O)N
	c1(C[C@@H](C(=O)O)N)ccccc1
	c1ccccc1C[C@H](N)C(O)=O

Table 2-7. RDKit and Open Babel generate two distinct canonical SMILESs

Software	The canonical SMILES
RDKit	N[C@@H](Cc1ccccc1)C(=O)O
Open Babel	N[C@H](C(=O)O)Cc1ccccc1

2.3.2.3 RNN (Recurrent Neural Network)

RNN solves sequential problems by training neurons receiving not only from the input layer but also from the output layer of the previous time step.[62,68] The information from the output layer of the previous time step is the memory, with which the MLP has no mechanism to deal. Figure 2-5 depicts the architecture of RNNs. The network, 'R', reads input from 'X' and output from the previous 'R' and generates 'Y'. All previous information is serialized and merged to the next neuron.

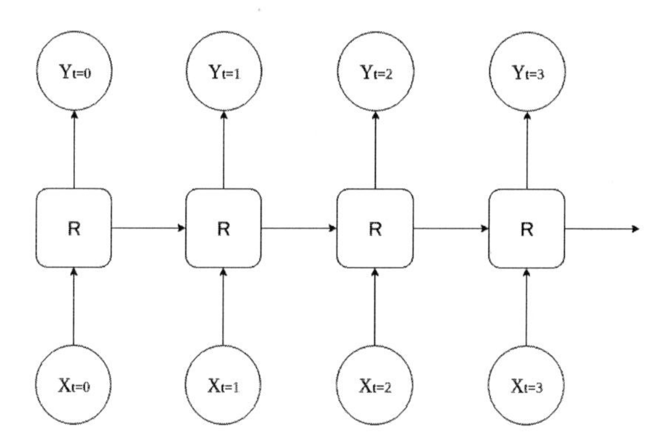

Figure 2-5. Schematic depiction of an RNN architecture.

Although an RNN integrates memory to infer the next state, long-term and short-term memory depend on the same variable. The short-term memory refers to previous states which are next to or nearby the current neuron, while the long-term memory comes far away from it. To make this simple, assuming a neuron is a multiplication operation, and Table 2-8 shows $Y_{t=0}$ to $Y_{t=3}$. In $Y_{t=3}$, the previous state, $X_{t=2}$, is multiplied by the square of R, and $X_{t=0}$ is multiplied by the fourth power of R. Both of these memories rely on R.

Table 2-8. RNN: The distance from previous states to the current neuron influences the memory effect.

Cell Output	Calculation
$Y_{t=0}$	$R \cdot X_{t=0}$
$Y_{t=1}$	$R \cdot (X_{t=1} + R \cdot X_{t=0})$
$Y_{t=2}$	$R \cdot (X_{t=2} + R \cdot (X_{t=1} + R \cdot X_{t=0}))$
$Y_{t=3}$	$R \cdot (X_{t=3} + R \cdot (X_{t=2} + R \cdot (X_{t=1} + R \cdot X_{t=0})))$

In order to fine-tune long-term and short-term memory separately, LSTM is proposed. There is an additional cell state in LSTM to regulate memory flow. Memory can flow through several neurons without any changes as long as it is desired. In order to manipulate the memory stream, there are mainly three gates to complete the task: forget gate, input gate, and output gate. These gates receive different driving strength from the input, previous state, and cell state, and are trained to control memory to increase accuracy.

2.3.2.4 Challenge

The synthesis prediction accuracy reaches 80.3% in the filtered USPTO dataset,[49] where 38648 reactions are considered, using the LSTM model with the attention mechanism. Although the model exhibits a higher accuracy, its mechanism is not satisfied by seasoned chemists. They argue that it is difficult to compare and analogy reaction transformation to language translation. Therefore, how to interpret the result is one key challenge to adopt this model.

The model is built to predict a single product, and all multiple-product reactions are filtered out as false cases. When considering multiple products, the permutation among them leads to an order dilemma for the sequential output. The question, which product will be the first output and which one will be the last, leads to a different sequence. Therefore, a translation model dealing with multiple products is another challenge.

The translation model is a non-template model. Authors assert that it has the potential to learn chemistry insight and solve reactions with unseen templates. However, there is no direct evidence to verify this assertion, and the analysis for unseen templates is needed to confirm or overthrow this.

2.3.3 Graph Model

2.3.3.1 Concept

Molecular connectivity can be expressed by the graph, which is constituted of vertices and edges. Vertices are an analogy to atoms, while edges denote to bonds. Synthesis prediction using the graph model has the following organizations: input features are atom properties and/or bond properties, and the output are atom edit configurations.[41] A molecular fragment editing provides more flexibility than reaction templates since a new template can be a different combination of fragment editing.

The graph model portrays the spatial relationship and linkages among atoms without specifying their exact coordinate positions. The graph model is a coordinate invariant where translation and rotation have no influence on this model. The graph representation of phenylalanine is illustrated in Figure 2-6, where atoms are labeled as atom symbol with a number, and the bond is shown in 'b' and tag number. Hydrogen atoms are omitted here.

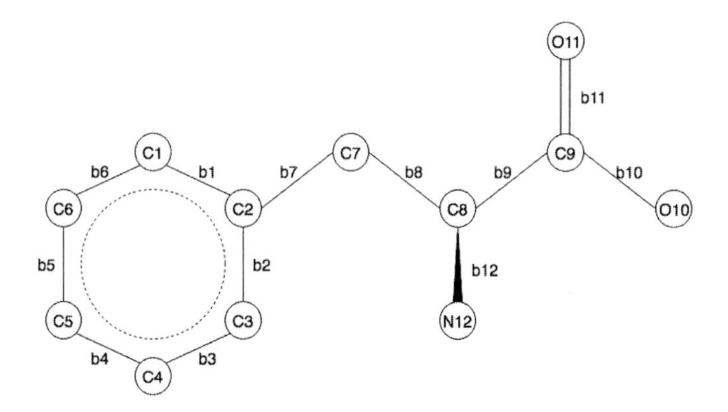

Figure 2-6. The phenylalanine graph.

2.3.3.2 Adjacency Matrix

The mathematical form of the graph is an adjacency matrix.[72] The atom adjacency matrix of phenylalanine is a 12x12 matrix in Figure 2-7. Zeroes are no connection between atoms while ones denote two bonded atoms. Meanwhile, Figure 2-8 describes the bond adjacency matrix.

If a matrix element is non-zero, two atoms are connected by it. Moreover, here is the bond type notation: 1=single, 2=double, 3=triple, 4=aromatic, 5=stereo-bond (pointing outward) and 6=stereo-bond (pointing inward). In the real computation, each element should be represented by a one-hot encoding, and it is a tensor with a dimension of 3.

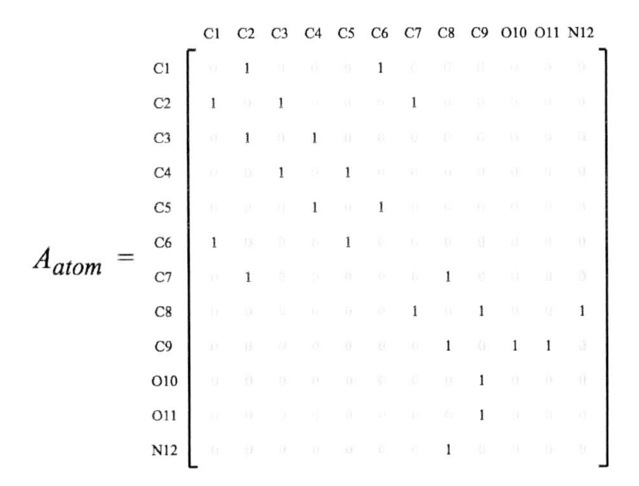

$A_{atom} =$

	C1	C2	C3	C4	C5	C6	C7	C8	C9	O10	O11	N12
C1	0	1	0	0	0	1	0	0	0	0	0	0
C2	1	0	1	0	0	0	1	0	0	0	0	0
C3	0	1	0	1	0	0	0	0	0	0	0	0
C4	0	0	1	0	1	0	0	0	0	0	0	0
C5	0	0	0	1	0	1	0	0	0	0	0	0
C6	1	0	0	0	1	0	0	0	0	0	0	0
C7	0	1	0	0	0	0	0	1	0	0	0	0
C8	0	0	0	0	0	0	1	0	1	0	0	1
C9	0	0	0	0	0	0	0	1	0	1	1	0
O10	0	0	0	0	0	0	0	0	1	0	0	0
O11	0	0	0	0	0	0	0	0	1	0	0	0
N12	0	0	0	0	0	0	0	1	0	0	0	0

Figure 2-7. The atom adjacency matrix of phenylalanine.

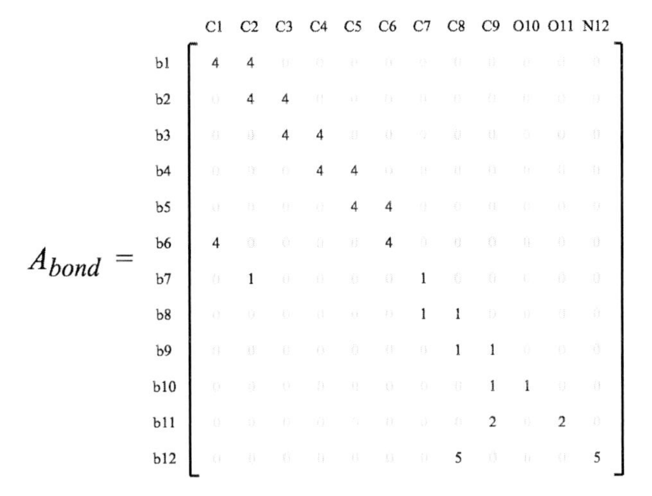

$A_{bond} =$

	C1	C2	C3	C4	C5	C6	C7	C8	C9	O10	O11	N12
b1	4	4	0	0	0	0	0	0	0	0	0	0
b2	0	4	4	0	0	0	0	0	0	0	0	0
b3	0	0	4	4	0	0	0	0	0	0	0	0
b4	0	0	0	4	4	0	0	0	0	0	0	0
b5	0	0	0	0	4	4	0	0	0	0	0	0
b6	4	0	0	0	0	4	0	0	0	0	0	0
b7	0	1	0	0	0	0	1	0	0	0	0	0
b8	0	0	0	0	0	0	1	1	0	0	0	0
b9	0	0	0	0	0	0	0	1	1	0	0	0
b10	0	0	0	0	0	0	0	0	1	1	0	0
b11	0	0	0	0	0	0	0	0	2	0	2	0
b12	0	0	0	0	0	0	0	5	0	0	0	5

Figure 2-8. The bond adjacency matrix of phenylalanine.

2.3.3.3 GCN (Graph Convolutional Neural network)

GCN extends coverage beyond non-Euclidean space.[73-76] CNN (Convolutional Neural Network) has a natural limitation: input must be described in a grid system.[62] Pixels on a 2D image plane and voxels in a 3D body are cases of CNN. However, systems, like group relationships or word vectors, are non-Euclidean space, which is challenging to be present in a grid system. In this scenario, GCN comes into play. GCN considers connection among elements.

By passing input features via an adjacency matrix, neighbors' information will aggregate on to the target atom, and this mechanism is message passing. There are different types of message passing, and, for example, the attention mechanism is also adopted to aggregate it. Although fingerprints also define local structure in a Boolean form indicating the presence or absence of functions, using a neural network to encode local graph generates vectors that give more parameters than one-bit Boolean. QSAR predictions based on GCN do already show state-of-the-art performance.[77] Therefore, this is the motivation to use the graph model in reaction predictions.

2.3.3.4 Challenge

In the filtered USPTO dataset with 38648 reactions, the synthesis prediction accuracy using the graph model is 74.0%.[64] This performance is higher than the template model but lower than the translation model. Many variations of GCN are proposed, which would update the message passing mechanism to adjust local information. It would be interesting to test these methods. The graph model is also not satisfying for seasoned chemists as that of the translation model. Therefore, how to interpret the result is one key challenge to adopt this model.

Connecting or breaking a bond is identified by enumerating through editing fragments. This method dissects the template into smaller fragments, and different combinations of fragments could build a new template. It overcomes the template limitation, but predefined fragment constraints still exist. Analyzing new reactions by templates and fragments is a way to understand how much advantage is added by the model.

There is no retrosynthetic prediction using the graph model. This solution requires a review of atom and/or bond changing mechanism from editing inference. Fusing multiple models is also possible by using different encodings at input and output terminals that could leverage the best encoding mechanism among the existing models.

2.4 Reporting

A reporting function is essential for ELNs since it provides a way to transfer, publish, and archive the research outcome. Nowadays, there is no absolute standard to format reports in the field of chemistry. An absolute standard would increase efficiency and accuracy of knowledge transfer since there is no clear-defined rule for presenting research-data. Manually taking care of detail formatting is also burdensome. As a result, a report function equipped with formatting assistance is beneficial, and this could foster the evolution of a new standard.

Although existing ELNs have a report function, the typical output format is PDF, which does not meet FAIR data standards: findability, accessibility, interoperability, and reusability.[25,78] FAIR data standards are highly desirable for digitalized data management. If a PDF file renders text in an image, it is not possible to search and find keywords on it directly. OCR (Optical Character Recognition) is adopted to reconstruct images back to raw text, but the accuracy is not perfect.[79] For PDF files containing text data, these are available for keyword searching, but it is challenging to interoperate the meaning of words. Additional algorithms are required to parse information from sparse text. As a result, reports with both original and structural text are desired. Embedding information into XML-like files can achieve this target.

ChemDraw[80] is a crucial standard for chemistry and biology research. Users can embed ChemDraw structures into Docx files by drag-and-drop. Moreover, editing the embed object will be updated to the document. No open-sourced tool can achieve this manipulation. A reporting function with this feature could reduce workload for scientists.

In summary, formatting assistance, XML-like structure, and ChemDraw embedding are vital features for a suitable report function. Designing an ELN system integrating these is a goal, and the resulting platform will be attractive to researchers and scientists.

3 Aim

To the current date, successful digitalization strategies in synthetic chemistry are still missing and the whole community suffers from missing data and scientific work that is hardly re-usable and reproducible.

The Chemotion-ELN was designed and is developed to assist users who work in the field of synthetic chemistry research to overcome this situation. It serves as a digital platform for the documentation of academic research. The ELN was designed in a way to increase the direct benefit to scientists who use it which should demonstrate the advantages of digital data management. The implementation of advantageous algorithms, mechanisms, and routines is of high importance for the successful strategy, in particular for academia, where the use of digitalization instruments is not yet mandatory and depends only on the popularity of the functionality.

The aim of this work is, therefore, to add functions to the Chemotion-ELN, which lead to substantial value for scientists, motivating them to work on digital systems continuously. One of the challenges of this thesis was to identify and implement those features that create incentives for users by applying the tools.

After investigation of work practices, routines and main bottlenecks of the Chemotion-ELN, three aspects were identified to be crucial: (1) chemical spectra editing and verifications, (2) reaction template and temperature predictions, and (3) reporting standards and templates.

First, spectroscopic data is an essential experimental readout to identify the molecular structure and purity of samples. It is required to determine whether a desired or expected compound is part of a given either pure sample or a mixture of compounds. However, nowadays, this identification and verification of chemical structures is done manually and only partially with the help of the machine in few cases.[17,18,20-23,36,39] In this work, deep learning and computer-assisted algorithms should be developed to support scientists to recognize and characterize chemical compounds to accelerate the analytical processes.

Second, the synthesis of molecules relies on synthetic route planning based on knowledge from accumulated experiences or databases. A system predicting reaction templates and environmental parameters, which aims to reduce the necessary resource using on exploring synthesis experiment, is long-awaited. In order to fulfill this goal, deep learning applications for template and temperature predictions should be analyzed in this work.

In the end, digitalized data on servers should be available for others in order to transfer knowledge. Additionally, an easy-to-generate supporting information is required to deliver content as publishing results. The idea of FAIR data should be adopted to maintain an economic research data management system. In order to achieve this vision, a reporting function should be built on the Chemotion-ELN, which could summarize selected data and deliver content in a findable, accessible, interoperable, and reusable manner. Therefore, generated reports should have raw data directly readable and searchable by both machine and human without additional transformation, e.g., optical character recognition methods.

By starting from the next chapter, these aspects are explained in detail.

4 Main Part - ChemSpectra

In almost all applications of synthetic chemistry, spectroscopic data is an essential tool to identify molecules and to determine purity. The output of spectroscopic measurements, named spectra, are important analytical results providing molecular evidence, like existence and proportionality, revealing whether desired structures or functional groups are present in the experimental outcome or not. Spectra can be obtained, e.g., from IR (infrared) and NMR (Nuclear Magnetic Resonance) measurements; the MS (mass) spectrometry also gives an output presented in a similar manner. The user can evaluate the obtained spectra by several commercial or open-sourced software products. However, to extract the key information from spectra, which requires time-consuming training and practice, is not a simple task.

Tools for the analysis of spectroscopic data exist but do not fulfill all requirements from chemists. On the one hand, open-sourced tools have limited functionality, e.g., only a few allow peak picking. Moreover, they are not integrated into a management system. On the other hand, commercial products are outstanding and allow advance interpretation. However, they are expensive, in many cases which cannot be integrated into management systems and cause dependencies that are not desirable in an academic environment. Both commercial and open-sourced tools have no all-in-one solution, i.e., one software provides the analysis of all IR, NMR, and MS spectroscopic data. Accordingly, an open-sourced tool, which solves these problems, is needed.

In this chapter, the ChemSpectra, a solution to the analysis of spectroscopic data, which is not comprehensive but allows a first full evaluation of three types spectra, is addressed. This open-sourced tool offers not only interactive user control, but also computer-assisted and deep learning evaluations. The management system, the Chemotion-ELN, keeps the modified spectra and the extracted information to achieve seamless data management. With all state-of-the-art functionalities, the ChemSpectra, which realizes automatically quality control and data curation of synthetic chemistry, enables a virtual environment of spectroscopic data.

4.1 Overview

ChemSpectra is a software solution providing a spectrum editor with computer-assisted and deep learning analysis tools. The system was designed with different modules (listed in Table 4-1). It integrates solutions for IR, MS, and one-dimensional ^{13}C and ^{1}H spectroscopy. Additionally, in the future, it will be possible to extend the use of the ChemSpectra to other spectra, e.g., RAMAN spectra. Primary users are people who work on synthetic chemistry.

Table 4-1. Modules of ChemSpectra. FE is frontend, the presentation layer. BE is backend, the data processing layer.

Module Name	Role	Function
react-spectra-editor	FE	Spectrum editor, Analyses editor
chem-spectra-client	FE	File management, Status notification
chem-spectra-app	BE	File conversion, Image generation, Peak picking, Third party communication proxy, Cheminformatics operations
chem-spectra-deep-ir	BE	IR functional groups existence inference

Figure 4-1 illustrates a high-level overview of the standalone system which can work independently without additional modules. The react-spectra-editor and the chem-spectra-client are frontend interface developed in JavaScript[81] on a React.js[82] framework offering a user-friendly single page application for uploading, reading, editing, and downloading spectra. The chem-spectra-app is a backend server developed in Python[83] on a Flask[84] framework providing data processing and cheminformatics operations. It is also the central proxy for other services, e.g., NMRShiftDB[36] and the chem-spectra-deep-ir service.

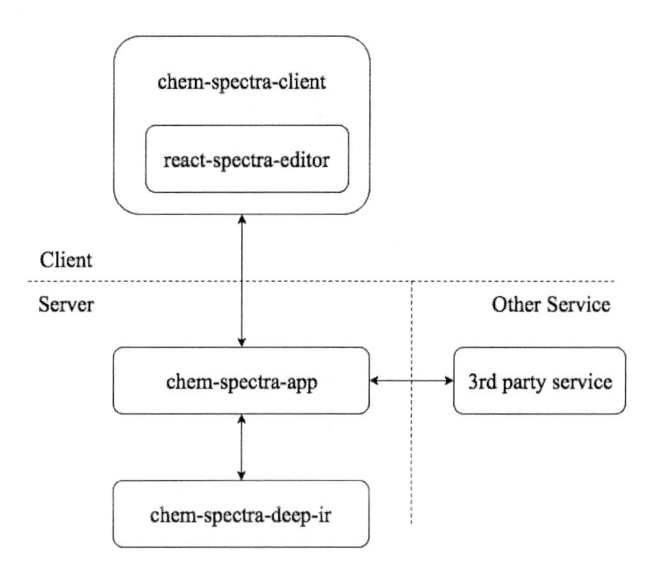

Figure 4-1. Overview of the ChemSpectra standalone system.

The chem-spectra-deep-ir verifies the existence of functional groups of a target molecule from an IR spectrum using the one-dimension convolutional neural network which was trained, validated, and tested by around five thousand spectra. This module is developed in Python on Tensorflow deep learning framework.

By default, there is no data storage function in the standalone version of ChemSpectra. Integrating ChemSpectra to the Chemotion-ELN can achieve storage management, where the user can inspect sample data in one place without accessing other applications.

In the Chemotion-ELN, DataCollector[10] sends spectra to the ELN after receiving them from the instrument. Manually transmitting files by email or portable devices is no longer needed. Then, spectra can be assigned to corresponding targets with drag-and-drop on the ELN interface directly. In the frontend, chem-spectra-client is no longer required, because the Chemotion-ELN, itself, takes care of uploading, downloading, and notifications. This design is an all-in-one data pipeline, and Figure 4-2 illustrates the architecture.

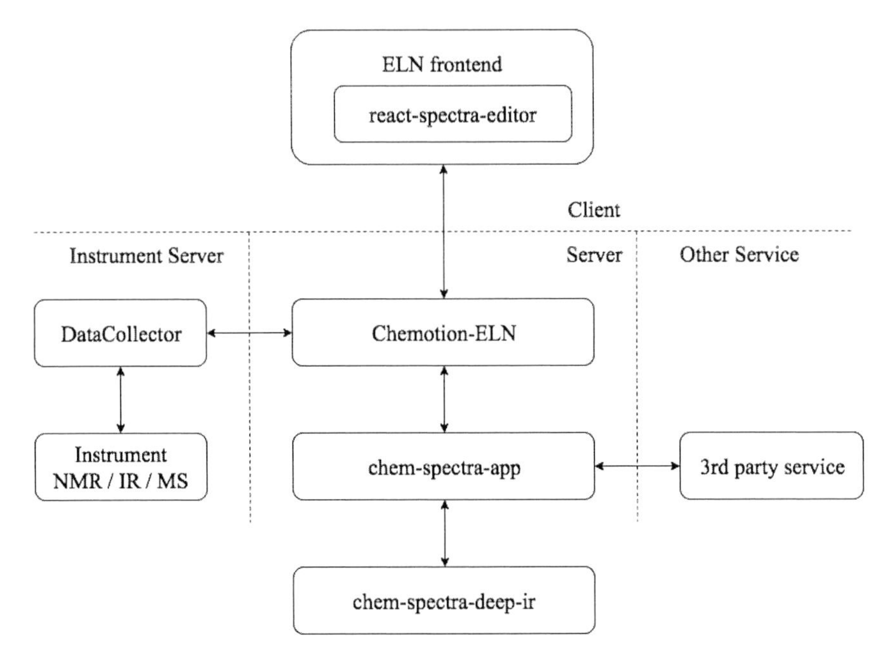

Figure 4-2. Overview of the ChemSpectra & the Chemotion-ELN integration.

4.2 React-spectra-editor

As a frontend editor, the react-spectra-editor is developed in JavaScript on a React.js framework offering a single page application. In order to render spectroscopic data into dynamic content, D3.js library[85], which creates, updates, and deletes SVG[86] elements efficiently by inspecting data difference without constant full reloading, is adopted. This implementation can reduce client system loading if there is one feature changed among a signal having thousands of points. This editor consists of two pages: “spectrum” and “analysis”.

4.2.1 Spectrum Page

In the spectrum page, three types of spectra, IR, NMR, and MS are displayed in two styles: the line plot and the bar graph. Data points in IR or NMR spectra are equally spaced and connected into a continuous curve in the line plot, in which a peak constitutes of one or several points. In contrast, data points of MS are discrete signals, unequally spaced, as the bar graph. Figure 4-3,

Figure 4-4, and Figure 4-5 show line-plot interfaces for IR, ^{1}H, and ^{13}C NMRs, while Figure 4-6 illustrates a bar graph of MS.

For all types of visualized spectra, the presentation is divided into two sections: the upper part is the primary editor for detail visualization and manipulation, while the bottom part is an overview panel in which the user can select a specific area (zoom in/out and shift focus function). Red identifiers mark features of the waveform. In the case of the NMR spectra, a green marker highlights the solvent peak. The threshold line (dotted green horizontal line) can be set and adapted to remove undesired peaks. Initial thresholds of IR, NMR, and MS, are 93%, 0.5% and 5% respectively, which are defined to avoid filtering out key features and to prevent including many noise peaks. Chemists in our group verify these values. For IR spectra, peaks above the threshold are removed, for all other types of spectra, the spikes under the threshold line are deleted.

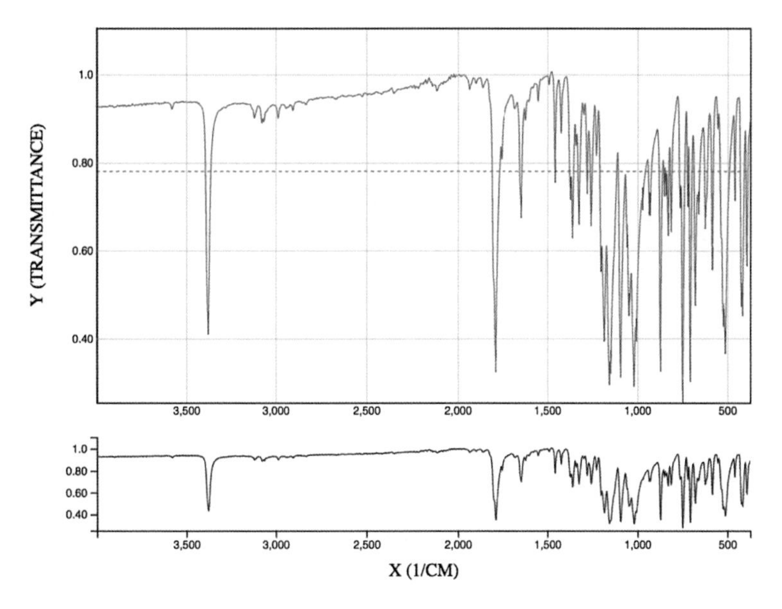

Figure 4-3. An IR spectrum view.

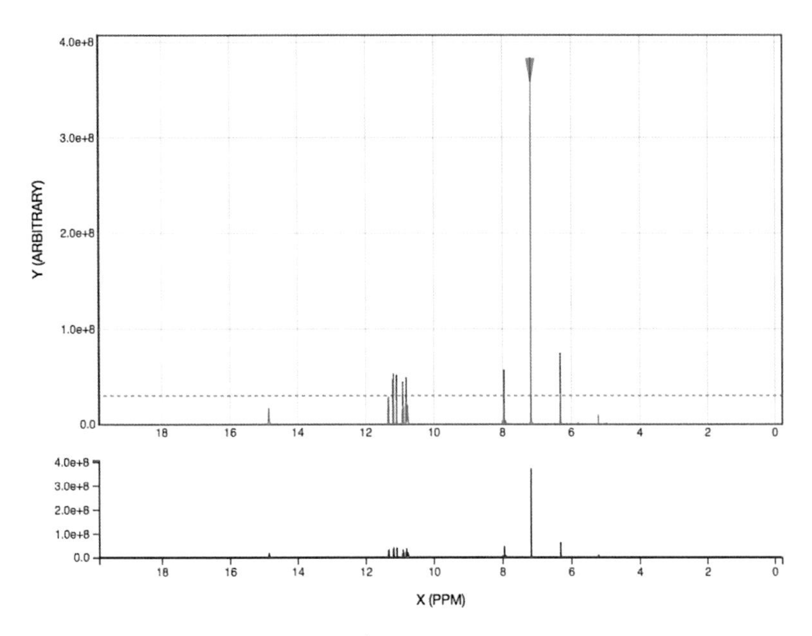

Figure 4-4. A ^{1}H NMR spectrum view.

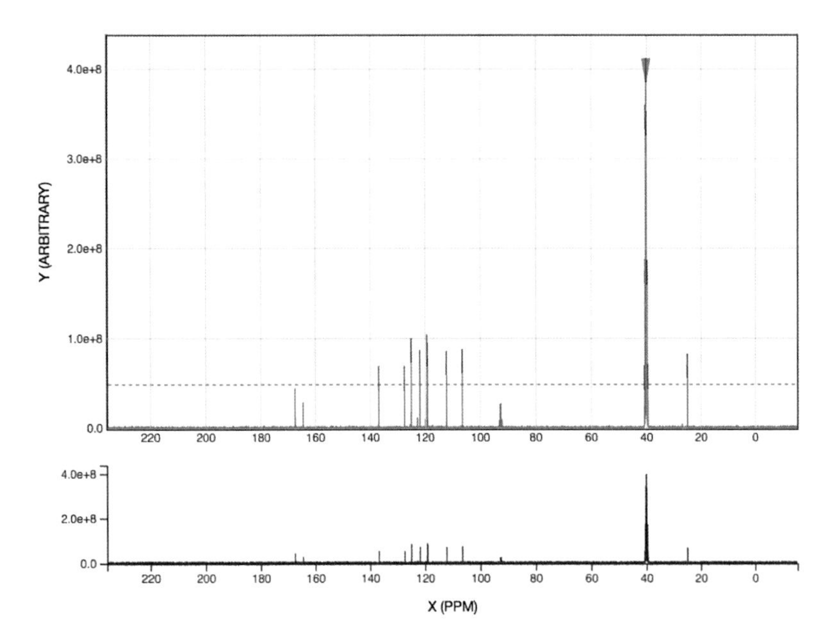

Figure 4-5. A ^{13}C NMR spectrum view.

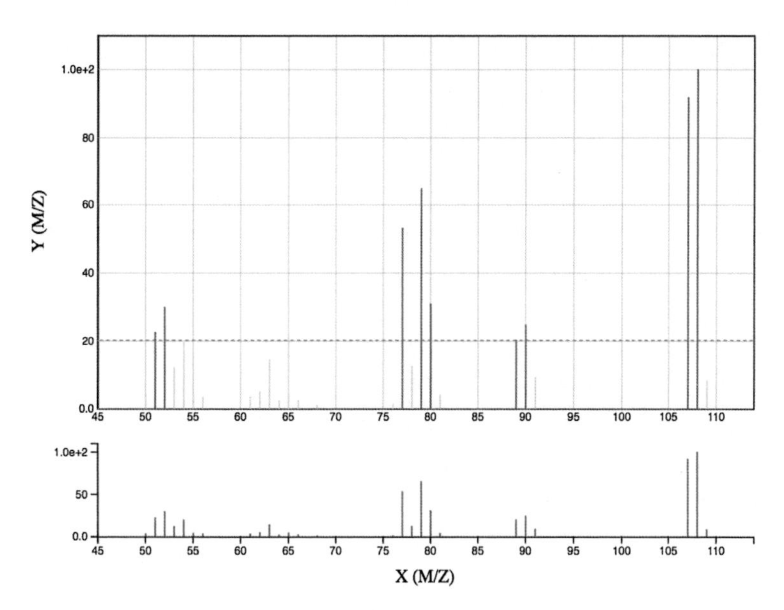

Figure 4-6. MS spectrum as a bar graph.

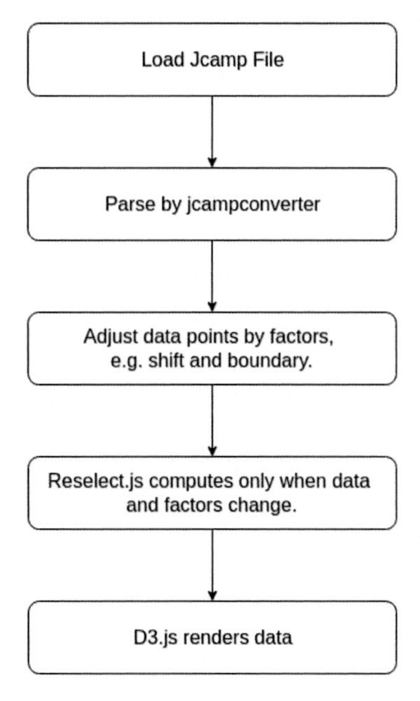

Figure 4-7. Data input and process flow.

4.2.2 Data Input

Figure 4-7 illustrates the data input flow of the react-spectra-editor. The data transferring between frontend and backend are JCAMP-DX files, which are extracted by jcampconverter,[87] a JavaScript library parsing the JCAMP-DX format. In order to parse NMR DEPT spectra and read JCAMP XYA tables, defined in the JCAMP-DX specification,[90] additional modifications are made and merged to the main branch.

There are around 1000 - 60000 points for typical spectra. Each point has two values corresponding to x and y positions. Re-rendering all points consumes many resources, and it could lead to bad user experience because of slow response. Therefore, a reselect library,[88] avoiding unnecessary computation, is added as a data container that maintains derived data efficiently by recalculating until data changes.

4.2.3 Control Panel

The control panel is on the right column of the page, where the user can execute actions (see Figure 4-8 as an example). There are three modes in the first section: set shift, add peak, and remove peak. These modes decide what is executed when clicking on the main editor. The second section lists settings like layout, solvent, threshold, and actions. Finally, there are sections for created and deleted peaks, where the user can review or redo peak editing.

The layout can be one type among IR, MS, ^{1}H, and ^{13}C, which is extracted from parsed parameters automatically. If the user decides that the layout is wrong, it can be changed manually. As to the solvent selection, the user needs to select which solvent is used in the measurement. This feature, only available for NMR, is not be displayed in IR and MS since there is no solvent reference for them. The threshold value can be input by the user or restored to the previous value. By toggling the icon, the user can retrieve the last peak-picking and settings. Action buttons include the following essential functions: save file, write peaks, and predict. The writing order of peaks is assigned by ascending or descending order.

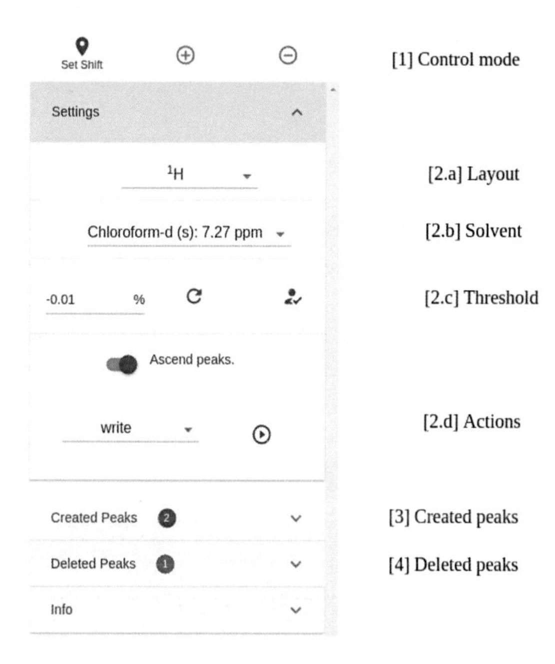

Figure 4-8. Control panel of the React-spectra-editor.

4.2.4 Editing in the Main Editor

Viewing local features can be approached by following steps: (1) selecting an area, e.g., a group of peaks, by dragging the mouse on the overview panel while pressing the left button; (2) scrolling to zoom-in or zoom-out; (3) local view adjustments by dragging the mouse while pressing the left button on the main editor. Restoring the main editor to the original scale is done by clicking on the right mouse button.

When there is a peak, filtered out by the threshold, without being labeled by the server, the user can switch to the "add peak" mode and click it in the main editor. A red marker is added and is appended to the "Created Peaks" list. The user can switch to "remove peak" mode and click on a red marker to remove it. All edited peaks are also displayed in "Created Peaks" and "Deleted Peaks" lists, in which the user can undo these by clicking corresponding labels.

In order to locate where the reference is for ^{1}H or ^{13}C NMR, the user must choose the "Set Shift" mode. By selecting a peak, the marker turns from red to green as its role changes. After selecting a corresponding solvent for the spectrum, the peak is shifted to the chosen value. NMR spectrum types (^{1}H vs ^{13}C) determine possible solvents and shifts.

4.2.5 Analysis Panel

In the analysis panel, a simulated result to verify the given spectroscopic data is displayed. By this tool, the user can analyze spectra with computer-assisted and deep learning algorithms, allowing to judge whether sub-structures or selected functional groups of the target molecule exist in the sample or not. An analysis is generated for IR, ^{1}H, or ^{13}C NMR by invoking the "predict" action.

4.2.5.1 IR Analysis

Deep learning models are trained to identify whether functional groups exist or not from massive IR spectroscopic data. This system reduces the burden of the traditional flow in which functional groups are recognized manually based on given absorption frequencies. Moreover, models in this thesis infer the specific functional groups, rather than general functional groups, e.g., alkanes and alkenes, that were studied before. The identification process, which is pure data-driven without manual encoding of chemical properties, enables the automatic quality assurance in synthetic chemistry.

The IR analysis panel lists functional groups and status. For inference of an IR spectrum, all functional groups are listed in SMARTS accompanying with a molecular SVG image, generated by RDKit, in which these groups are highlighted in different colors for more comfortable visual perception. The machine status is the inference, which can be "accept", "reject", or "unknown", from the deep learning server. Only trained models with high accuracy, equal to or higher than 80%, are included in the server. The "unknown" status means that there is no good model available for that structure. When the status is "reject", the model infers that the functional group is not in the sample based on the analysis of the obtained experimental spectrum. Table 4-2 presents the status vs. the condition.

Table 4-2. IR analysis status vs. condition.

Status	Condition
Accept	The functional group exists.
Reject	The functional group does not exist.
Unknown	The system does not have a suitable model for this functional group, and the user has to determine this manually.

The accuracy of the test set, computed by the deep learning model, is the machine confidence. If it is very close to 100%, the user should pay close attention to examples with rejected status since the interpretation for functional groups having high confidence is likely to be right (and therefore the result is wrong with a high probability). The owner of the analysis can verify the status manually by adding an interpretation in the field "owner". This editing is useful when the user does not agree with the inferred outcome or determines manually as the functional group is "unknown" to the system. The owner's interpretation has the highest priority for an overarching interpretation of the analysis result. Figure 4-9 illustrates an example of the IR analysis.

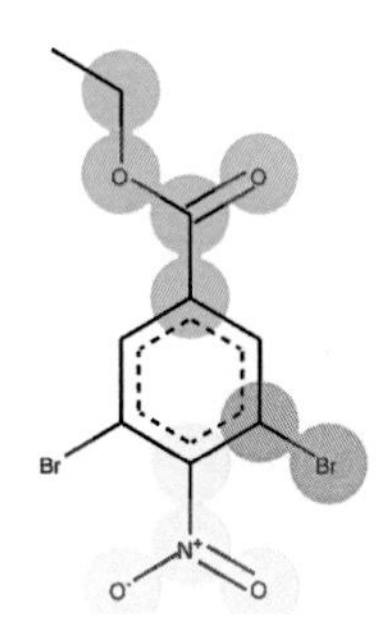

	FG SMARTS	Machine Confidence	Machine	Owner
1	c-,:[N&+](=O)-,:[O&-]	99.19 %		
2	c-,:C(=O)-,:O-,:C	95.20 %		
3	c-,:[Br]	88.46 %		

Figure 4-9. Analysis panel for an IR spectrum.

4.2.5.2 NMR Analysis

The NMR analysis tool gives the different shifts of the peaks in the spectrum that were obtained experimentally and a simulation that was gained based on the structure (with the help of

NMRshiftDB calculation).[36] Each peak corresponds to one target atom: the target for ^{1}H is a proton, and the carbon atom is for ^{13}C NMR. In the analysis table, each row has an atom number which follows the order of atoms in the molfiles. The level of difference leads to four statuses: "accept", "warning", "reject" and "missing". Table 4-3 lists the condition. Figure 4-10 is an example of the ^{13}C NMR analysis.

Table 4-3. NMR analysis status vs. condition.

Status	Condition
Accept	Simulation and experiment peaks match.
Warning	The difference is a little bit higher, but still in an acceptable range.
Reject	The difference is too high, and the local environment does not match to the simulation result.
Missing	There is no matching peak.

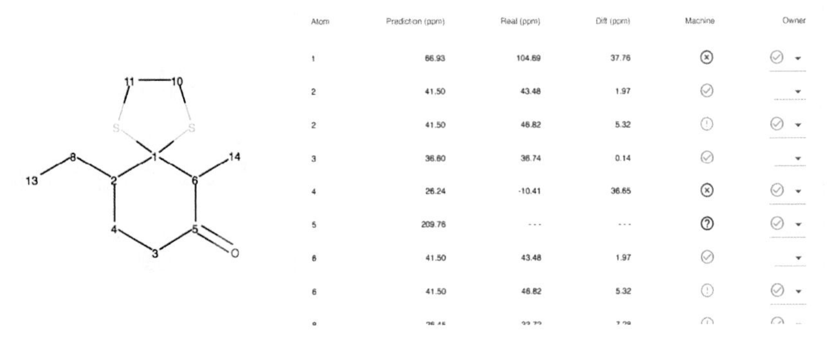

Atom	Prediction (ppm)	Real (ppm)	Diff (ppm)	Machine	Owner
1	66.93	104.69	37.76		
2	41.50	43.48	1.97		
2	41.50	46.82	5.32		
3	36.60	36.74	0.14		
4	26.24	-10.41	36.65		
5	209.76	---	---		
6	41.50	43.48	1.97		
6	41.50	46.82	5.32		

Figure 4-10. Analysis panel for a ^{13}C NMR spectrum.

4.3 Chem-spectra-client

The chem-spectra-client is only needed for the standalone service when the integration to the Chemotion-ELN server is not present. It has three main functions: communication to the backend server, file management, and notifications.

The chem-spectra-client hosts a webpack server and communicates with the backend server. A browser can load a static asset from it at a defined port (port 3006 as default). The backend system, the chem-spectra-app, acts as an API server occupying at default port 2412. When the user clicks buttons like saving data in the user interface, the API server is responsible for the file processing. In order to accomplish the communication, the browser always sends requests to the webpack development server, which is the only service exposed to the public. For operations relating to the API server, the webpack development server is the proxy to bypass the request to port 2412 (illustrated in Figure 4-11).

Figure 4-12 depicts the chem-spectra-clients user interface. There are three rows: (1) file management, (2) react-spectra-editor, and (3) notifications and output.

In the file management row, three files can be uploaded, and two of them are required. The Molfile and spectrum files are compulsory, while the predicted result is optional. The necessary chemical information, e.g., molecular mass, of the molecule, is extracted from the uploaded Molfile.

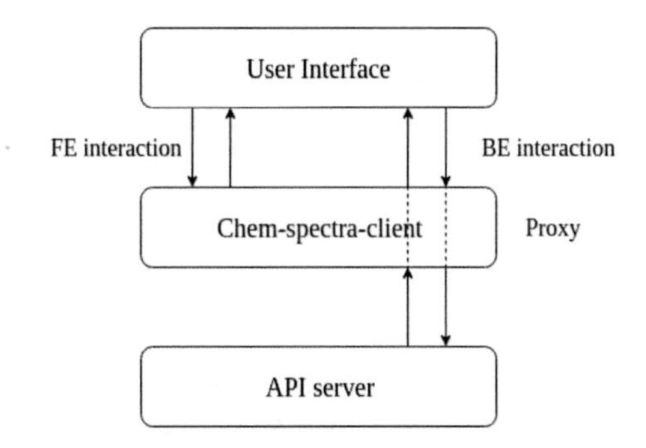

Figure 4-11. The chem-spectra-client as the proxy to the API server.

Figure 4-12. The chem-spectra-clients user interface.

The system accepts only the following file extensions for the spectrum file: jcamp, jdx, dx, mzml, and raw files. Mzml and raw files are formats only for MS spectra. Jcamp, jdx, and dx formats can carry IR, NMR, and MS spectra. The predicted result is recorded in a JSON file, containing inference outcome. The maximum file size is 10Mb.

The uploaded files are sent to the API server for further processing, and the returned response contains a processed jcamp file. Then, the jcampconverter inside the react-spectra-editor can extract information and can render it on the page. Asynchronous operations trigger a loading icon as the page is waiting for the response. After an operation is finished, a notification message pops up to inform about the status. It displays an error message for debugging when there is something wrong.

4.4 Chem-spectra-app

In Table 4-4, there are three main functions of the chem-spectra-app. The app is a Flask API server which controls routings and communicates with third party services. Algorithms, NMR and, IR inference, which consumes high computing resources, are separated from this server.

Table 4-4. Functions of the chem-spectra-app.

Function	Examples
spectra transformation	1. Spectra files decoding and composing, peak-picking, and image generation
spectra inference	1. NMR shift identification 2. IR functional group verification 3. Selection of the best MS scan
cheminformatics conversion	1. Conversion of molfiles to molecular mass 2. Identification of functional groups of a molecule

Table 4-5. Possible input spectrum formats for the chem-spectra-app.

File format	Extension	Applicable types
JCAMP-DX	.jcamp, .jdx, and dx	NMR, IR, and MS
mzML	.mzml	MS
raw	.raw	MS

When integrating with the ELN server, the chem-spectra-app is a satellite server in charge of all spectra-related processing excluding storage and management. In this scenario, the ELN regulates main requests.

4.4.1 Spectra Transformation

The chem-spectra-app can process three possible input formats: JCAMP-DX, mzML, and raw files. Table 4-5 lists the accepted formats, filename extensions, and spectrum types. Several classes are built to take care of different file extensions and spectra types (see Figure 4-13).

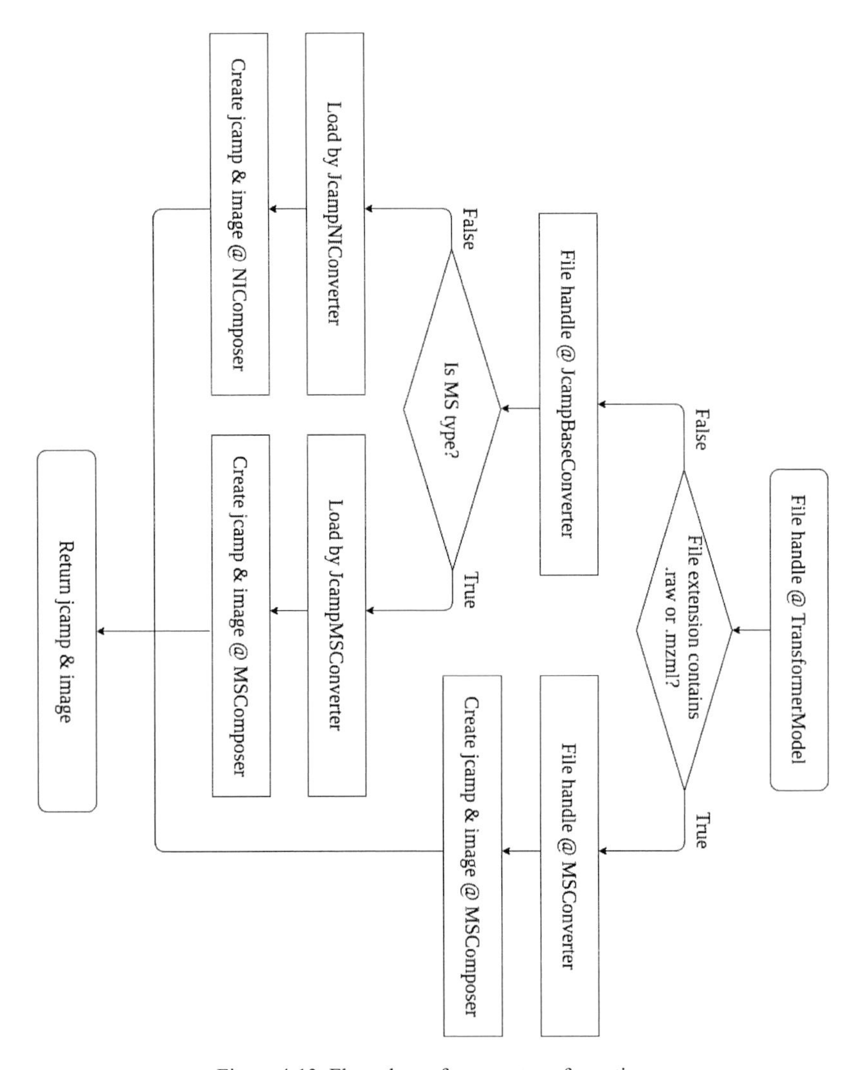

Figure 4-13. Flow chart of spectra transformation.

The given file extension and attributes lead to the determination of the provided type. If the file extension is .mzml or .raw, MS-related classes handle file parsing and output composition. For a JCAMP-DX file, it could be a NMR, IR, or MS file. The JcampBaseConverter class reads the type from file content using the nmrglue library,[89] which is explained in detail in the next section. Then, the file is passed to different classes. In the end, the output format for all spectra is unified to the JCAMP-DX format, with the extension '.jdx'.

4.4.1.1 Parsing NMR and IR Files

In the chem-spectra-app, the JCAMP-DX format is read by nmrglue,[89] a python library which is capable of manipulating NMR spectra across several formats, including JCAMP-DX, Bruker, NMRpipe, Sparky, and Varian/Agilent. It is a JCAMP-DX parser in this application. Based on JCAMP-DX specifications for IR, NMR and MS spectra,[90] several predefined fields provide information of JCAMP-DX files. In order to know the spectrum type, the "DATATYPE" field contains one keyword among "NMR SPECTRUM", "NMRSPECTRUM", "INFRARED SPECTRUM", and "MASS SPECTRUM". The first three keywords identify files to be NMR and IR spectra, while designed python classes for MS spectra, present in the next section, process files containing the last keyword, "MASS SPECTRUM".

After reading a file, a python object keeps parameters and data points, including auto-picked and manual-picked peaks. If there are no existing auto-picked peaks, the scipy library[91] identifies the peaks automatically. Since the user might set a new reference which induces lateral movement, this is also reflected in auto-picked peaks. When new user-defined peaks are present in the parameters, they replace the previously given manual-picked part. Figure 4-14 shows a complete flow chart.

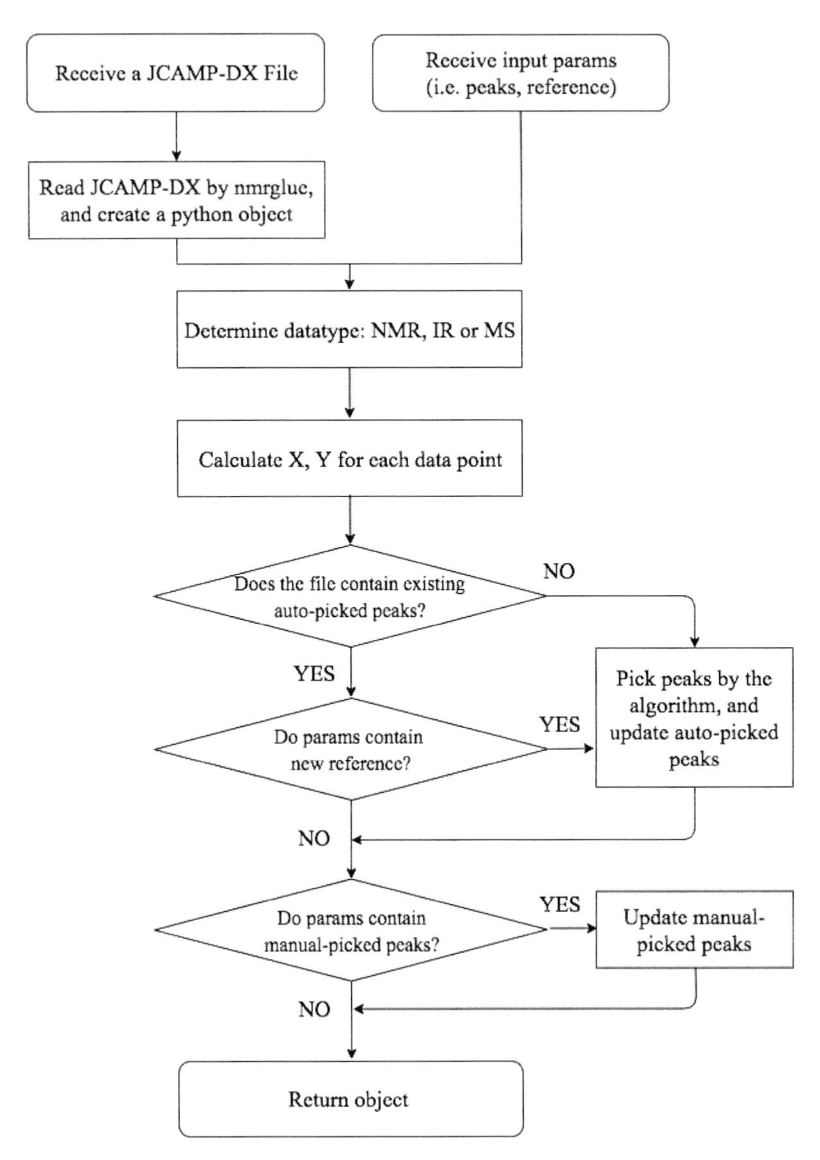

Figure 4-14. Flow chart of reading a JCAMP-DX file.

4.4.1.2 Parsing MS Files

In order to unify the processing of a spectra, the system converts all files to the JCAMP-DX format. Since MS spectra can have all three possible formats, e.g. JCAMP-DX, mzML, and raw, it must be transformed into JCAMP-DX first (see Figure 4-15).

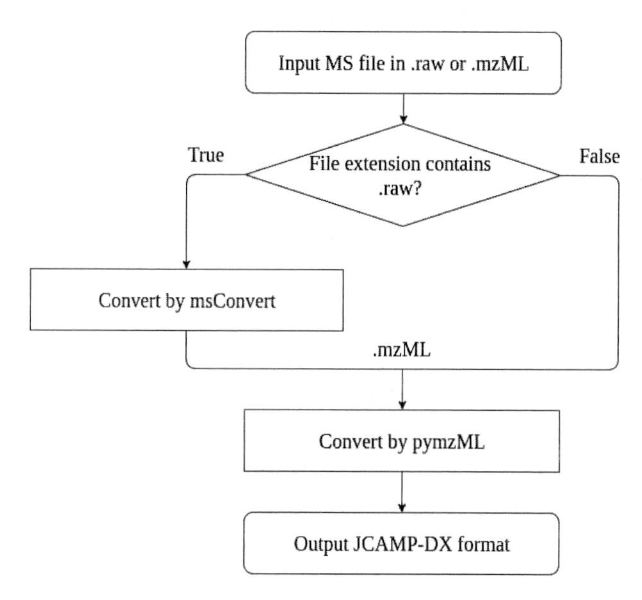

Figure 4-15. Mzml and raw conversion to JCAMP-DX inside the python MSConvert class.

Mzml and raw formats are dedicated to MS spectra. Mzml is an open-sourced format that is similar to the human-readable XML tree structure. Because the raw format is a binary data that is not human readable, it has to be decoded before further processing. For this purpose, msConvert in Proteowizard[92] is employed to convert MS files from raw to mzML. In the application, msConvert in a Proteowizard docker container which is called by the MSConvert python class achieves this job.[93] Then, mzML files are converted to JCAMP-DX using pymzML,[94] an open-sourced python mass spectrometry file parser.

A MS spectrum file contains multiple scans, of which an algorithm is integrated to select the best one automatically. The details of this process will be explained later. After the identification of the best scan number, data points, parameters, and the threshold, a JCAMP-DX file keeps them for fast information retrieval in the future. Based on the different possible requests, the JCAMP-DX file is sent back to the client side or stored in the Chemotion-ELN server. Meanwhile, if an image of the spectrum is needed, it is generated by the python matplotlib library.[95]

When the input file is a JCAMP-DX file, the process is similar to that of NMR and IR spectra. The "DATATYPE" field is "MASS SPECTRUM" and it is parsed and composed using JcampMSConverter and MSComposer classes, respectively.

However, if the input format is not one of raw, mzML, or JCAMP-DX, the user needs to convert the file into one of them by other software. This workaround is the current limitation of the chem-spectra-app.

4.4.1.3 Composing JCAMP-DX Format

When the parsing, described in previous sections, is finalized, the next step is to compose a new JCAMP-DX file. Figure 4-16 illustrates this composition, which is an abstract of an NMR JCAMP-DX file. To sum it up, there are three inner blocks, these are data points, auto-picked peaks, and manually-picked peaks. An outer link block wraps these three inner blocks together.

The composed JCAMP-DX file only contains the necessary information. Users can open the same spectrum multiple times for editing and reviewing. This compact file reduces data transmission.

Reducing server loading is one of our guidelines when designing functions. During the planning phase, extracting files in the server and sending raw data points to the client side is also considered. Other services, like nmrpro,[29] adopt this pattern. However, the server has to extract and process files whenever a user opens the editor. This configuration takes more resource in the server. Contrary, in the ChemSpectra, sending a file to the client side, which is then decoded by jcampconverter, take less resource in the server.

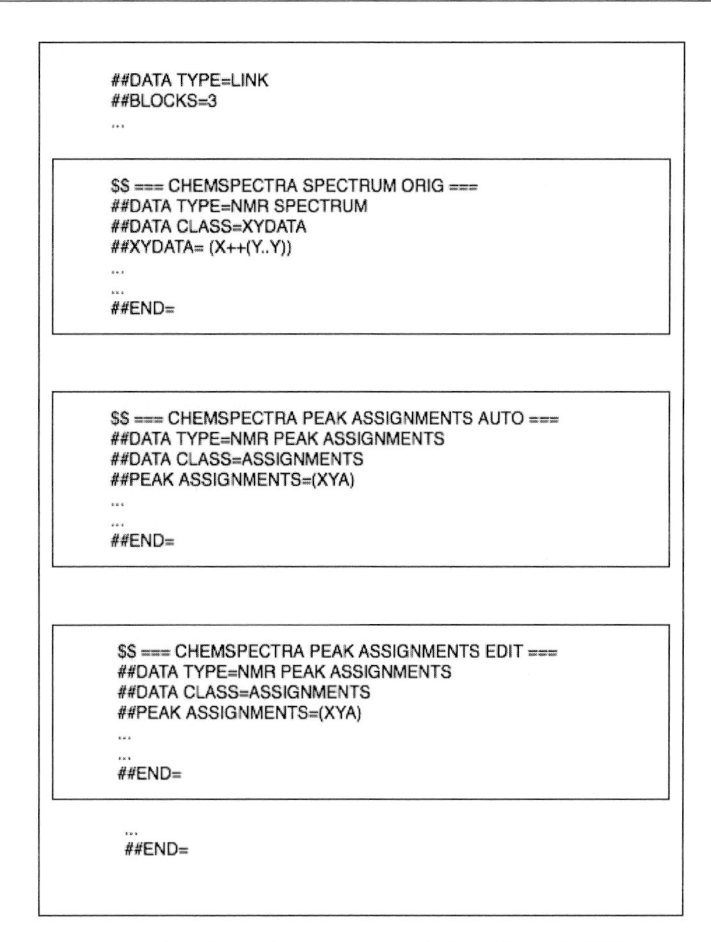

```
##DATA TYPE=LINK
##BLOCKS=3
...

$$ === CHEMSPECTRA SPECTRUM ORIG ===
##DATA TYPE=NMR SPECTRUM
##DATA CLASS=XYDATA
##XYDATA= (X++(Y..Y))
...
...
##END=

$$ === CHEMSPECTRA PEAK ASSIGNMENTS AUTO ===
##DATA TYPE=NMR PEAK ASSIGNMENTS
##DATA CLASS=ASSIGNMENTS
##PEAK ASSIGNMENTS=(XYA)
...
...
##END=

$$ === CHEMSPECTRA PEAK ASSIGNMENTS EDIT ===
##DATA TYPE=NMR PEAK ASSIGNMENTS
##DATA CLASS=ASSIGNMENTS
##PEAK ASSIGNMENTS=(XYA)
...
...
##END=

...
##END=
```

Figure 4-16. Abstraction of a composed NMR JCAMP-DX file.

4.4.2 Spectra Inference

Spectra are used for structure elucidation and as proof of the existence of molecular sub-structures. Currently, this verification process is done manually or with partial assistance of machines. The ChemSpectra project aims to provide computer assistance to reduce human loading when analyzing spectra. Table 4-6 lists the currently available software programs that were developed to provide inference services.

Table 4-6. Services for spectra analysis.

Spectrum type	Service
NMR	NMRShiftDB
IR	chem-spectra-deep-ir
MS	chem-spectra-app

In a MS plot, the x-axis gives the mass-to-charge ratio, while the y-axis gives the intensity of a detected signal. It is expected to have high intensity at the exact molecular weight of the material. Besides, it is also possible to find high intensity at exact molecular weight +1, +23 and +39, due to proton, sodium, and potassium adduct formation. Therefore, the intensity at these positions is compared with the highest intensity in a scan to get the ratio of the two intensities. The scan with the highest ratio is selected as the best scan because the target molecule is considered to be present in the sample. The chem-spectra-app automatically takes the scan with the highest ratio of the desired mass-to-charge value, and it avoids manual selection based on intuition.

As to NMR spectra, the quick-check service from NMRShiftDB[36] is leveraged. Its algorithm determines whether the chemical shifts match local structures of the target protons or carbons. The NMRShiftDB algorithm adopts HOSE codes to describe the atom environment by encoding multiple spheres of different radius in a SMILES-like format. By searching HOSE codes in a shift-to-HOSE-codes database based on experimental spectra, several shifts can be linked to one target atom. An idea shift value of an atom in a molecule is the average of these candidates.

If the difference between a real shift and a simulated shift is small, ^{13}C NMR within +/- 5 ppm, ^{1}H within +/- 1 ppm, its status is “accept”, and the spectrum supports the existence of sub-structure with high confidence. Medium deviation, ^{13}C NMR within +/- 10 ppm, ^{1}H within +/- 2 ppm, indicates lower confidence with the “warning” status. The status is “reject” when the deviation is more significant than previous definitions, and in this condition, the corresponding sub-structure cannot be approved by the spectrum. The status “unknown” implies that the algorithm cannot find any shift matching the structure.

Molecular vibration modes, which absorb energy related to local structure, are observed by IR spectra, which is a critical analysis for functional group recognition. A rule-based functional group identification[16] is not considered in this thesis to be a good solution due to an unavoidable band overlap. Moreover, it cannot suggest which specific functional groups are supported by the spectrum as the criteria take only a definition of generalized functional groups.

The chem-spectra-deep-ir adopts convolutional neural networks to recognize the presence of functional groups from the full spectrum profile, not discrete peaks. Detail of the training and evaluation of networks are discussed later. This method is data-driven without pre-encoded rules.

A review of computer-assisted and deep learning solutions integrated into the ChemSpectra project is listed in Table 4-7.

Table 4-7. Summary of computer-assisted solutions.

Spectrum type	Computer-assisted and deep learning solutions
NMR	Confirm the status of each target atom based on the difference between the simulated shift of molecular structure and the real spectrum shift.
IR	Confirm whether functional groups of the target molecule exist or not based on the spectrum profile.
MS	Pick the best scan based on desired mass-to-charge ratios.

4.5 Building IR Inference Models

In this section, the process to train and validate deep learning models for IR spectra is described. The goal is to build models inputting a spectrum waveform, and then the output is whether functional groups of the corresponding molecule are likely to exist or not.

4.5.1 Data Source

Training a deep learning model requires massive data which is, in case of IR spectra, hardly available due to missing open repositories offering a high number of IR data. The IR spectra used for the work described in this thesis are taken from two sources: Chemspider[96] and our in-house database. Chemspider is a chemical structure database offering data from commercial vendors and crowdsourcing, i.e., users can contribute to the database. IR spectra, downloaded from Chemspider around the mid of 2017, were parsed by nmrglue. A valid spectrum must meet two requirements: (1) it has to be accessible via nmrglue and (2) its SMILES have to be readable by RDKit to create a molecule object. Altogether, 4347 valid IR spectra from Chemspider and 884 from our in-house database were obtained and used.

4.5.2 Spectrum Preprocessing

Before feeding spectra into the model, the spectra are standardized by a series of processing steps as shown in Table 4-8.

Table 4-8. Preprocessing methods to standardize spectra.

Method	Explanation
Same type	All spectra are of type absorption.
Resampling	The spectrum profile is resampled. All data points locate at the integer wavenumber.
Padding	All spectra are standardized to cover values between 400 - 4000 cm^{-1}. Data points outside this range are cut off. If a spectrum does not record data nearby the boundary, they are extended from the closest points.
Normalization	Data points are normalized between 0 and 1.

First, all spectra must be the absorption type. Each peak indicates the absorption of energy by intramolecular vibration related to a certain functional group. Since sampling across different instruments is not always at the same wavenumbers, resampling is applied to extract data from the same locations while maintaining the same spectrum profile. Wavenumber at integer values between 400 - 4000 cm^{-1} are analyzed, and the interval is 1 cm^{-1}. For example, assuming a fragment with original sampling at 1100.8, 1101.4, 1102.0, 1102.6 and 1103.2 cm^{-1}, data at 1100, 1101, 1102 and 1103 cm^{-1} are interpolated without modifying the original profile.

Ending points from different instruments may vary. There are endings at around 400, 500 or 600 cm^{-1}. Padding additional points around the boundary are necessary to synchronize the input data without adding artificial features. For example, when the smallest wavenumber of a spectrum is given at 550 cm^{-1}, a point at 400 cm^{-1} is added with the same intensity. In order to keep the same interval, points between 400 cm^{-1} and 550 cm^{-1} with interval 1 is interpolated. This procedure is also applied in the same manner to endpoints on the other side.

In the end, the full span of intensity, i.e., y-direction, is rescaled to lay within 0 to 1. Maximum and minimum intensities of the original profile are extracted. All original points are normalized based on Equation 4-1.

$$y' = \frac{y - y_{min}}{y_{max} - y_{min}} \qquad \text{Equation 4-1}$$

Figure 4-17 is an example to demonstrate the procedure to reprocess a spectrum in the described manner. The IR spectrum of the molecule 3-methyl-4-penten-1-ol (CSID 110120, transmission spectrum) is downloaded from Chemspider and rendered by the python matplotlib library. The wavenumber span is 550 - 3846 cm^{-1}, and the intensity is between 0.0 to 0.014. Two red dotted lines on left and right sides are boundaries at 400 and 4000 cm^{-1}, respectively.

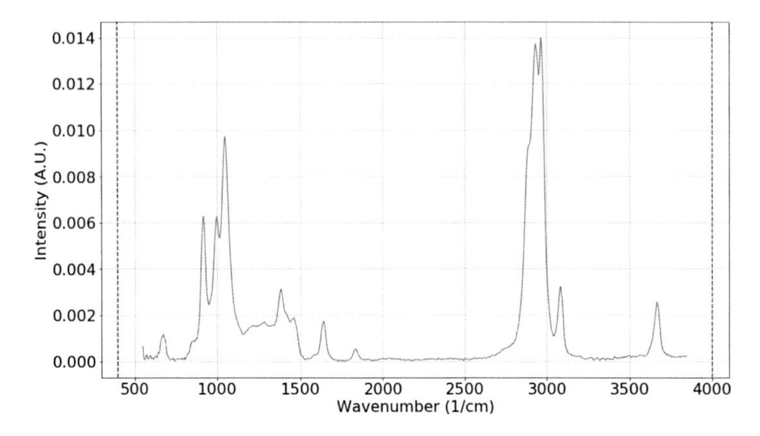

Figure 4-17. The original spectrum of Chemspider CSID 110120.

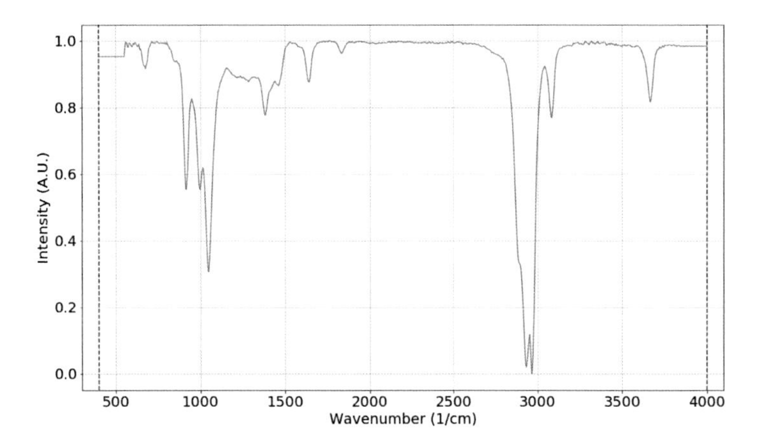

Figure 4-18. The processed spectrum of Chemspider CSID 110120.

The processed spectrum gained from the demo data is illustrated in Figure 4-18. It is flipped vertically to be an absorption spectrum. Left and right ending points match the defined range of 400 - 4000 cm^{-1}, where intensity is confined between 1.0 and 0.0. All data points are integral values. The most important fact of all is that the spectrum keeps the same profile as given in the original source.

4.5.3 Functional Group Extraction

Functional groups are extracted by an algorithm[97] which is implemented in the RDKit IFG function.[98] The algorithm is described in Table 4-9 based on a publication of ERTL *et al.*[97] The method used in RDKit IFG searches targets by specific atom features. By inputting a molecular SMILES into the RDKit IFG function, an array of functional group SMARTS is generated.

The functional groups of phenylalanine are displayed here for a clear explanation in Figure 4-19. This method identifies two groups: (1) C-,:C(=O)-,:O, and (2) C-,:N. In the molecule, there are three heteroatoms: two oxygen at 10, 11, and one nitrogen at 12. Table 4-10 contains the explanation of the constitution of these two groups.

Table 4-9. Atom features considered as functional groups.[97]

Criteria
1. All atoms that are not carbon or proton (→ heteroatoms)
2. Carbons with following features: - acetal carbons - carbons connected to heteroatoms via non-aromatic double or triple bond - carbons in nonaromatic c-c double or triple bonds - carbons in oxirane, aziridine, and thiirane rings
3. Functional groups are fragments containing previous targets with radius-one neighbor carbons included.

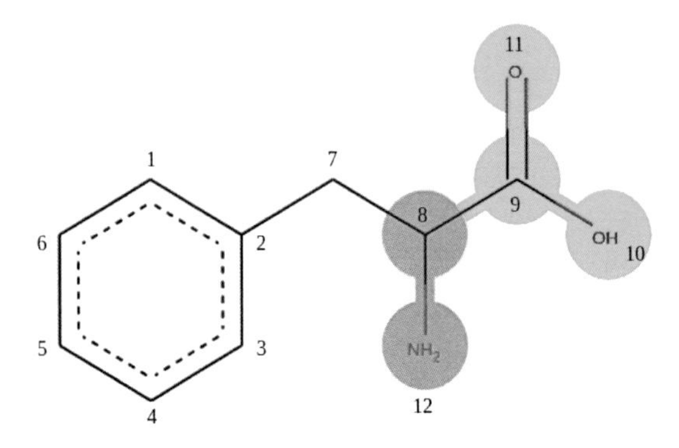

Figure 4-19. Phenylalanine functional groups are highlighted by SMARTS matching.

Table 4-10. Explanation of phenylalanine functional group formation.

Functional group	**Explanation**
C-,:C(=O)-,:O	Carbon 9 connects to oxygen 11 via a double bond. It is also the neighbor of oxygen 10. Besides, carbon 8 is the radius-one neighbor of this group. Therefore, this functional group has atoms: 8, 9, 10, and 11.
C-,:N	The carbon 8 is a radius-one neighbor to the nitrogen 12. They form the functional group.

4.5.4 Deep Learning Models

It is crucial to determine which deep neural network should be used to achieve the goal. Desired networks should predict the existence of functional groups from a given spectra.

Each input spectrum contains 3600 floats, which an additional encoder is not necessary to apply on. Moreover, the spectrum waveform is a one-dimensional array. It is logic to use MLP (Multi-Layer Perceptron) or 1D CNN (Convolutional Neural Network) to capture features relating to different wavenumbers or local distribution. An RNN (Recurrent neural network) is not applicable, because data is not sequential.

The output is the existence of functional groups, and one molecule can have zero or more functional groups. They are not mutually exclusive. By combining functional groups, it can also be reduced to a multi-class classification problem.[39] For example, classes can be (1) alkane, (2) alcohol, (3) both alkane and alcohol. However, it is difficult to scale this model when the number of functional groups is high. Therefore, it is reasonable to combine several binary classification models.

Because spectra are collected from real experiments, and the number of supporting spectra for each functional group is different. The RDKit IFG algorithm inspects in total 5231 molecules. Chapter 10.1 lists the top 44 frequent functional groups. An imbalanced dataset is perceived. The most frequent functional group has 666 positive spectra (12.7% from 5231 spectra), while the 44th frequent one has only 40 supporting spectra (<1% from 5328 spectra). It is challenging to evaluate multi-label classification from this imbalanced dataset. Therefore, combining several binary classifications is adopted in this thesis, and multi-label classification will be explored in the future.

4.5.5 Model Training and Verification

The prediction was made by a binary classification model per one functional group owing to an imbalanced dataset. Therefore, predictions for 44 functional groups need 44 binary classifiers. Each classifier can be evaluated legitimately based on its own positive and negative samples without bias from other functional groups.

Training, verification, and testing datasets are split into a ratio: 67:22:11 for each functional group through the whole dataset. Positive and negative samples for each functional group are split, respectively, and balanced accuracy is employed to evaluate the performance. Table 4-11 and Table 4-12 list two examples of sample split from the 1st and 44th frequent functional groups.

Table 4-11. 1st sample splits for training, verification, and testing.

C-,:O 666 out of 5231 spectra have this functional group.

Training		**Validation**		**Testing**	
positive	**negative**	**positive**	**negative**	**positive**	**negative**
446	3058	147	1009	73	498

Table 4-12. 44th sample splits for training, verification, and testing.

C-,:I 40 out of 5231 spectra have this functional group.

Training		**Validation**		**Testing**	
positive	**negative**	**positive**	**negative**	**positive**	**negative**
26	3477	9	1148	5	566

The balanced accuracy is the average of respective accuracy for each class, which can solve interference and bias from the majority in the overall accuracy of an imbalanced dataset. Taking the training set of the 44th functional group, C-,:I, as an example, it means that there are 26 spectra in the training set containing this functional group, and 3477 spectra do not have it (only spectra in the training set are considered). If the model predicts that all are negative, the overall accuracy is 99.26%, which is a good value. However, balanced accuracy is only 50%, since all positive samples yield false predictions. As a result, balanced accuracy is a fair evaluation of our problem.

To sum up, each network for a functional group has input in dimension *[batch size, 3600]*, because there are 3600 points between 400 - 4000 cm^{-1}. There are two classes of output, as a binary classifier is generalized as a multi-class model with two classes, i.e., the spectrum has the functional group or no corresponding group is detected. Table 4-13 and Table 4-14 are model architectures of MLP and 1D-CNN models. Figure 4-20, Figure 4-21, Figure 4-22, and Figure 4-23 are loss and balanced accuracy curves for the 1st frequent functional group, C-,:O. All batch normalization layers have momentum 0.9, and the epoch count is 300.

Table 4-13. MLP model architecture.

Layer	Structure
Hidden 1	1. Fully connected layer with output dim = 4000. 2. Batch normalization layer 3. RELU activation layer
Hidden 2	1. Fully connected layer with output dim = 1200. 2. Batch normalization layer 3. RELU activation layer
Hidden 3	1. Fully connected layer with output dim = 512. 2. Batch normalization layer 3. RELU activation layer
Hidden 4	1. Fully connected layer with output dim = 128. 2. Batch normalization layer 3. RELU activation layer
Hidden 5	1. Fully connected layer with output dim = 32. 2. Batch normalization layer 3. RELU activation layer
Output	1. Fully connected layer with output dim = 2.
Loss	SOFTMAX
Optimization	Adam

Table 4-14. 1D-CNN model architecture. (results: Table 10-1)

Layer	Structure
Hidden 1	1. 1D-CNN with kernel size = 5, # of filters = 16 2. RELU activation layer 3. Batch normalization layer 4. 1D max pooling layer with size = 2
Hidden 2	1. 1D-CNN with kernel size = 5, # of filters = 32 2. RELU activation layer 3. Batch normalization layer 4. 1D max pooling layer with size = 2
Hidden 3	1. 1D-CNN with kernel size = 5, # of filters = 64 2. RELU activation layer 3. Batch normalization layer 4. 1D max pooling layer with size = 2 5. Flatten layer
Hidden 4	1. Fully connected layer, with output dim = 128. 2. Batch normalization layer 3. RELU activation layer
Hidden 5	1. Fully connected layer, with output dim = 36. 2. Batch normalization layer 3. RELU activation layer
Hidden 6	1. Fully connected layer, with output dim = 8. 2. Batch normalization layer 3. RELU activation layer
Output	1. Fully connected layer, with output dim = 2.
Loss	SOFTMAX
Optimization	Adam

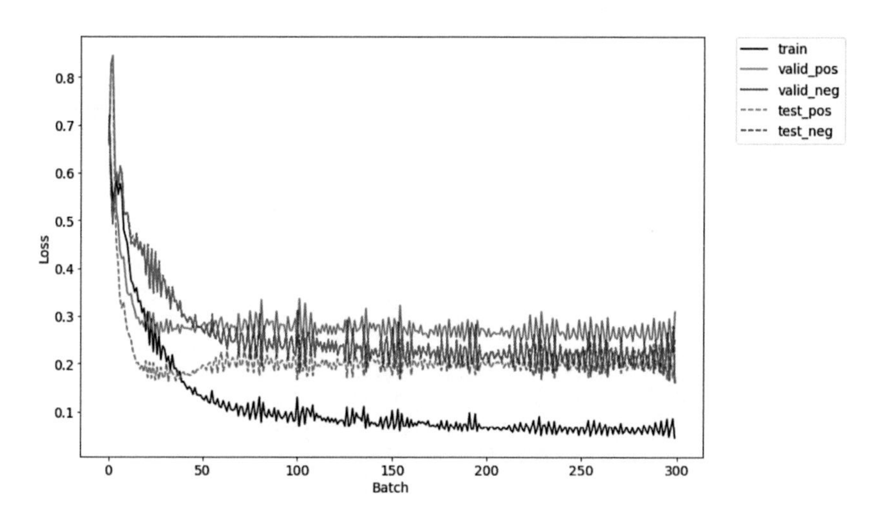

Figure 4-20. MLP loss for the 1st frequent functional group, C-,:O.

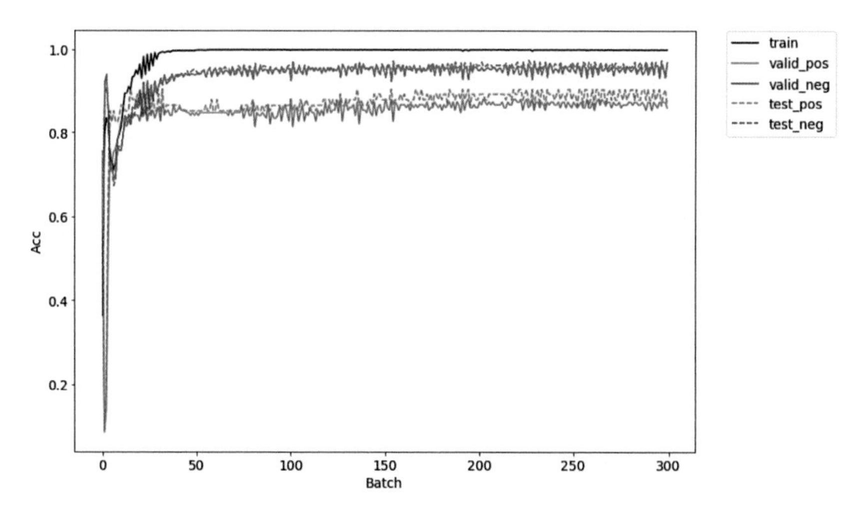

Figure 4-21. MLP balanced accuracy for the 1st frequent functional group, C-,:O.

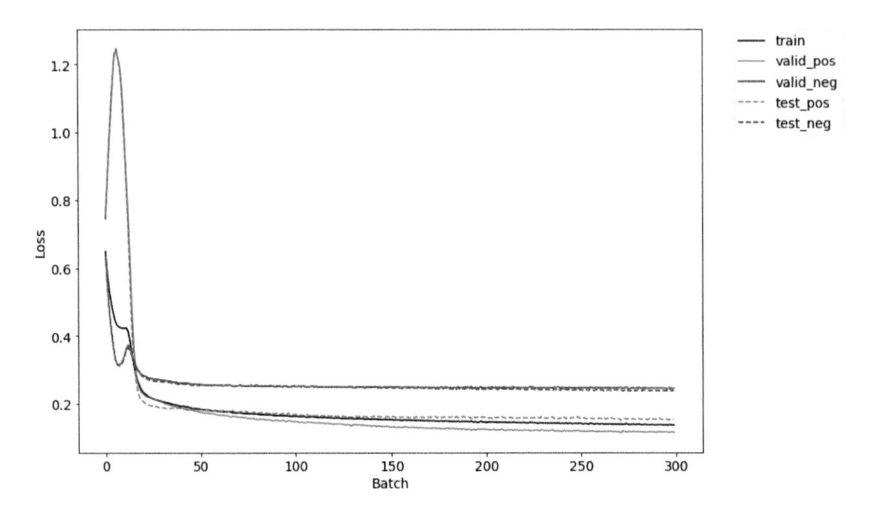

Figure 4-22. MLP loss for the 1st frequent functional group, C-,:O.

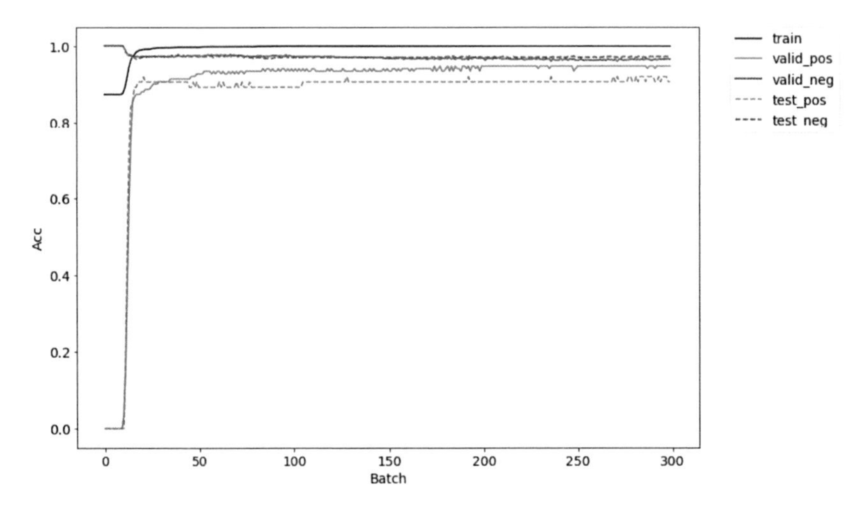

Figure 4-23. MLP balanced accuracy for the 1st frequent functional group, C-,:O.

MLP and 1D-CNN models perform similar balanced accuracy, but the CNN model shows a smoother curve with less noise. Therefore, 1D-CNN is adopted for the rest of the functional groups. Chapter 10.1 lists the result. For 36 out of 44 functional groups existence predictions can achieve the accuracy equal to or higher than 80%.

4.6 Chem-spectra-deep-ir

The chem-spectra-deep-ir is a Flask server responsible for IR inference. When the user wants to verify whether a spectrum contains all functional groups of a molecule, the first step is the data extraction in the chem-spectra-app. The functional groups of the molecule are extracted from the given Molfile, and spectroscopic data points are extracted and preprocessed using the IR JCAMP-DX file. After preprocessing, the chem-spectra-app sends the data to the chem-spectra-deep-ir. Then, the chem-spectra-deep-ir predicts the existence of functional groups using trained Tensorflow 1D CNN models. Figure 4-24 illustrates this flow.

The chem-spectra-deep-ir only takes care of predictions, while spectrum and cheminformatics calculation are done at the chem-spectra-app, which prevents the processing codes from being scattered into multiple sites.

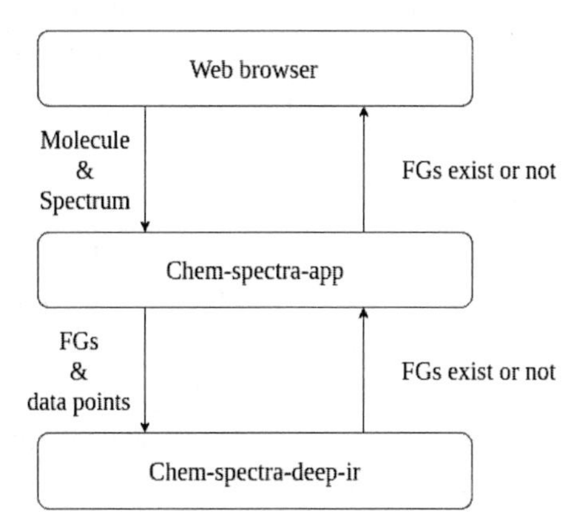

Figure 4-24. IR inference flow.

4.7 Chemotion-ELN Integration

The Chemotion-ELN integration is more complicated than that of the standalone server, because files generated from the instrument, data persistence, information sharing, and workflow management need to be considered.

In the standalone server, when the user closes the web page, files do no longer exist. The information exists only temporary in the server memory during the processing. The server discards the information as soon as the conversion finishes. To retrieve the edited data after closing the page, the user has to repeat the actions.

Concerning the ELN server, the original uploaded files and newly composed files are persisted for two purposes. First, original files are valuable data without modifications, and they are necessary when there is further discussion regarding data correctness. Second, composed files are kept to avoid repeating generations. Meanwhile, records inside the database are created and updated to track the procedure.

A state machine is adopted to orchestrate process managements in the Chemotion-ELN, because of the need for rigorous managements between old and new files. It is realized using AASM, a ruby library that extends finite state machines for Ruby classes. Every uploaded file is stored in the disk which is accompanied by a record in the database table, and the *aasm_state* column indicates at what state the file is (see Table 4-15).

After initialization, a file with a png extension is set to the image state. Files other than JCAMP-DX, mzML, raw and png extensions are set to the *non_jcamp* state. When a file has a spectrum extension, it is set as the *queueing* state waiting for further processing. Once the file is decoded and composed to a new file, the original file is set to the *done* or *backup* state, depending on creating or editing action, respectively. The composed file is also set to the *peaked* or *edited* state, for creating or editing action, respectively. Only spectrum files with the *peaked* or *edited* state can be viewed in the react-spectra-editor. The design pattern limits that there is only one viewable spectrum file in each analysis. If there is an error during conversion, the file is set to *failure* for a further review. Figure 4-25 illustrates state diagrams.

Table 4-15. State definitions for the attachment table.

AASM state	**Description**
idle	All files are initialized as this state.
queueing	A JCAMP-DX file is waiting for processing.
peaked	A composed JCAMP-DX file has automatically-picked peaks.
edited	A composed JCAMP-DX file has both automatically and manually picked peaks.
done	This state denotes an original JCAMP-DX file, and a new JCAMP-DX is composed based on this.
backup	A new JCAMP-DX is composed based on this file.
image	The file is an image file.
non_jcamp	The file is not a JCAMP-DX file.
failure	Something has gone wrong during the process.

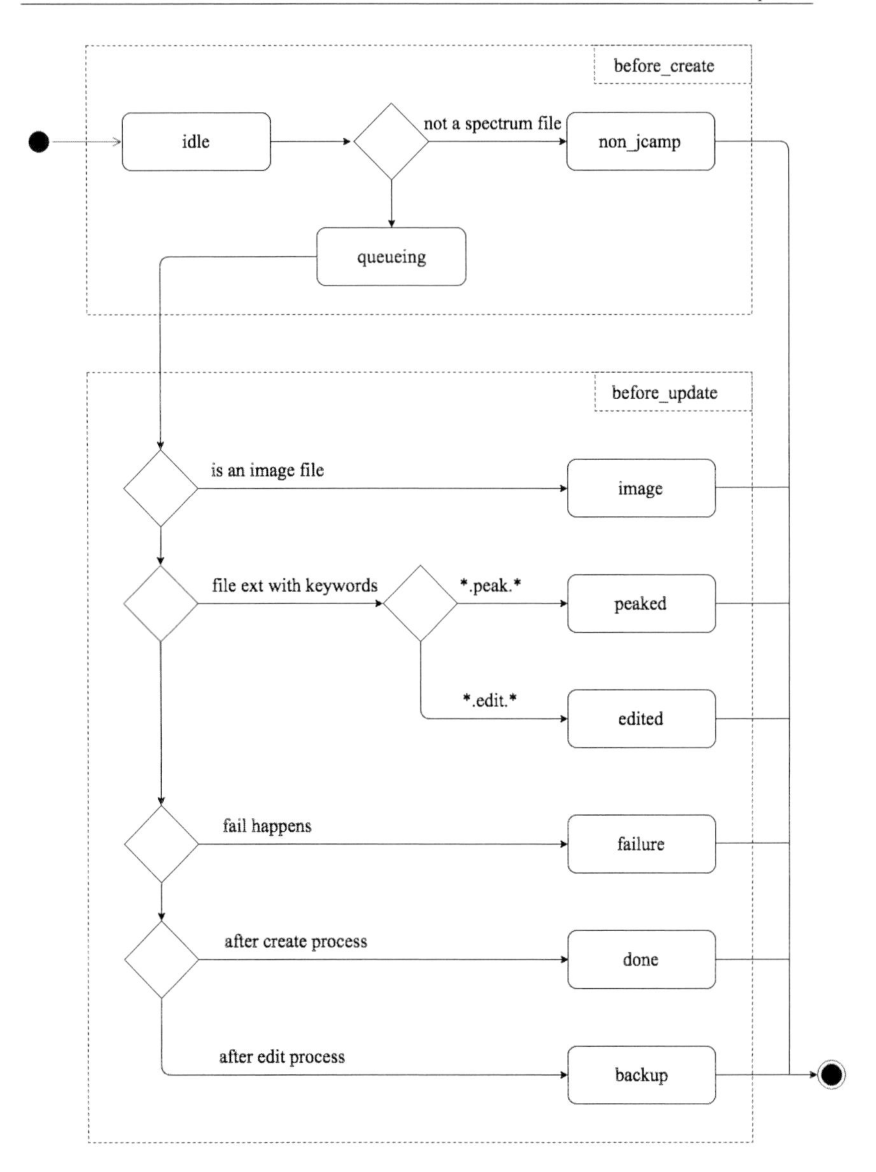

Figure 4-25. Sequence diagram of the state machine.

4.8 Quality Assurance – Quality Control and Quick Check

The tool "Quality assurance" provides the summary and final evaluation of IR, NMR, and MS analyses that can be collected and processed by the ELN. This function is only available in the Chemotion-ELN because the quality assurance tool needs different types of analyses of the samples to be analyzed and other open-sourced tools offering the necessary functions are not available. Currently, one standard is implemented, and in the future, more standards, which will be selectable by the users, will be included based on demand. This standard, which rules are designed by seasoned chemists in our group, is favorable for synthetic chemistry.

This section contains two parts. First, an overview of data availability, quick check, and quality control is rendered. Moreover, a score, which summaries all data by one number, gives the user the confidence level of combined analyses. Second, the analyses detail which gives the key features of all analyses is described.

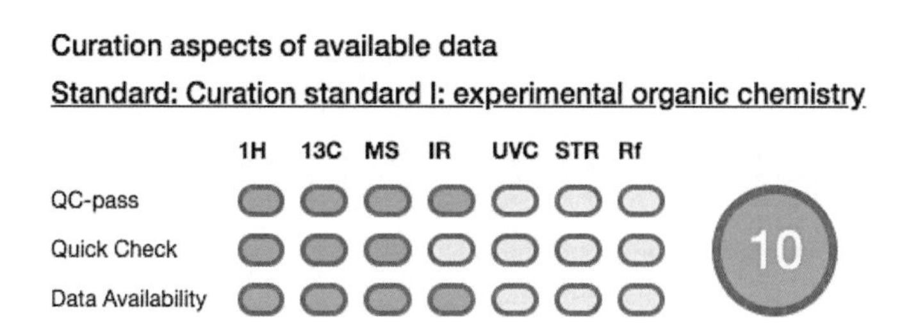

Figure 4-26. An example of an overview of the quality assurance tool in the ELN UI.

4.8.1 Overview

The overview part illustrates the result of three essential aspects, the data availability, quick check, and quality control for each spectrum type, as shown in Figure 4-26. Explanation of these aspects are listed as followings:

(1) "Data availability" indicates whether a valid spectroscopic file (in the format JCAMP or Raw as a specific application) is uploaded and processed by ChemSpectra.

(2) "Quick check" presents the result of the verification of text descriptions which were inputted or confirmed by the user. NMR and MS, but not IR, have a quick check function (as a result of the chemical interpretation habits). For NMR spectra, the quick check function checks whether the peak count in the given description matches the

target atom number. This function is enabled for proton (^{1}H) and carbon (^{13}C) NMR as a result of the given molecular sum formula. As to MS spectra, if the exact molecular weight or the weight of possible adducts with a proton, sodium-, or potassium-cation can be found in spectrum peaks, it is valid and illustrated in green color.

(3) "Quality control", abbreviated as QC-Pass in the user interface, is the spectroscopic analysis by computer-assisted and deep learning algorithms. As discussed in the previous sections, the owner can correct machine inference if a mistake happens.

The score value evaluates how much evidence is provided by the user to support the fact that the target molecule is likely to exist in the experiment result. There are two types of evidence: (1) text description of spectroscopic data, e.g., peak locations, no matter the corresponding file exists or not, and (2) spectrum files as ultimate approval for the accuracy of the data.

This experimental organic chemistry standard encourages using experimental results, by uploading spectrum files to the server, to support the presence of the target molecule and its sub-structure. Therefore, analyses, including spectrum files, have a higher score than that only confirmed by the text description. The score value is from -4 to 10, in which a higher value indicates a higher confidence. The calculation of the score is described as follows:

(1) Score = -4 – -1. There is at least one analysis in ^{13}C NMR, ^{1}H NMR, or MS which has only quick check but no spectrum files as evidence.
(2) Score = 0 – 3. Quick check and files are provided in all ^{13}C NMR, ^{1}H NMR, and MS analyses. However, two or three analyses have no result from computer-assisted or deep learning quality control.
(3) Score = 4 – 9. Quick check and files are provided in all ^{13}C NMR, ^{1}H NMR, and MS analyses. However, one analysis has no result from computer-assisted or deep learning quality control.
(4) Score = 7 – 10. Quick check and files are provided to all ^{13}C NMR, ^{1}H NMR, and MS. All analyses have positive computer-assisted or deep learning quality control. IR quality control is the key factor to determine the final score.

In all cases, the exact score is determined by (1) the correctness of ^{13}C NMR, ^{1}H NMR, and MS quick check and (2) IR, ^{13}C NMR, ^{1}H NMR computer-assisted or deep learning quality control. The score is set to zero when all quick checks are positive, but all quality control is negative.

UV, crystal structure, and R_f are reserved in the interface. They will be part of future work to improve the process based on additional analytical and experimental data.

4.8.2 Analysis Detail

The analysis detail part has four elements: ^{1}H NMR, ^{13}C NMR, MS, and IR.

In ^{1}H NMR and ^{13}C NMR, “according to the user” is the text description of peaks provided by the user. Peak count is compared to the target atom count deduced from the sum formula. In quality control, “Signals detected” are peaks identified in the spectroscopic data, while predicted peaks are based on the molecular structure from NMRShiftDB which are listed in “Signals detected (NMRShiftDB)”. Machine and owner evaluations are summarized. For the pass criterion, maximum one failure is allowed in the machine prediction; zero failure is allowed in the owner verification. Figure 4-27 and Figure 4-28 show the analysis detail of ^{1}H NMR and ^{13}C NMR.

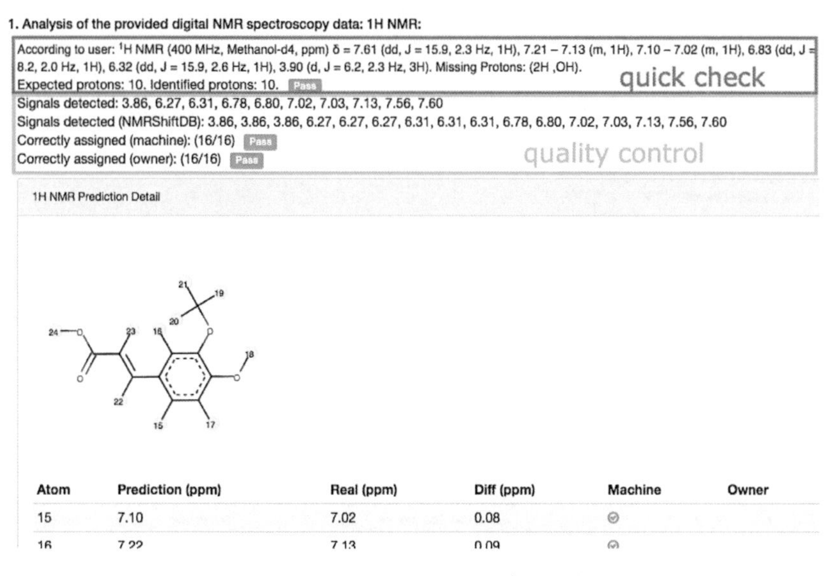

1. Analysis of the provided digital NMR spectroscopy data: 1H NMR:
According to user: ¹H NMR (400 MHz, Methanol-d4, ppm) δ = 7.61 (dd, J = 15.9, 2.3 Hz, 1H), 7.21 – 7.13 (m, 1H), 7.10 – 7.02 (m, 1H), 6.83 (dd, J = 8.2, 2.0 Hz, 1H), 6.32 (dd, J = 15.9, 2.6 Hz, 1H), 3.90 (d, J = 6.2, 2.3 Hz, 3H). Missing Protons: (2H ,OH).
Expected protons: 10. Identified protons: 10. Pass
quick check
Signals detected: 3.86, 6.27, 6.31, 6.78, 6.80, 7.02, 7.03, 7.13, 7.56, 7.60
Signals detected (NMRShiftDB): 3.86, 3.86, 3.86, 6.27, 6.27, 6.27, 6.31, 6.31, 6.31, 6.78, 6.80, 7.02, 7.03, 7.13, 7.56, 7.60
Correctly assigned (machine): (16/16) Pass
Correctly assigned (owner): (16/16) Pass
quality control
1H NMR Prediction Detail
21
19
20
24
O
23
16
O
18
O
O
22
15
17
Atom
Prediction (ppm)
Real (ppm)
Diff (ppm)
Machine
Owner
15
7.10
7.02
0.08

Figure 4-27. Analysis details of ^{1}H NMR.

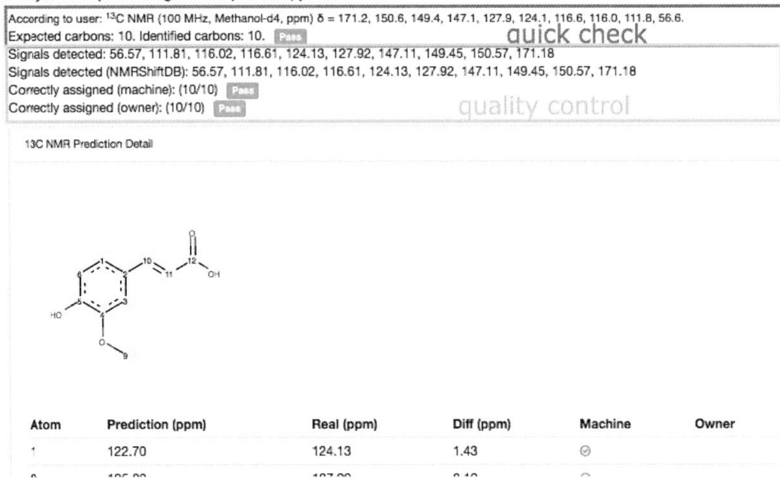

Figure 4-28. Analysis details of ^{13}C NMR.

3 Analysis of the provided digital mass spectrometry data:

Identified Mass peaks (*m/z*) = 177.05 (71%), 195.07 (100%), 196.07 (10%), 209.08 (27%), 282.28 (12%), 302.25 (10%), 304.26 (43%), 320.23 (38%)
Selected scan: 8.
Exact molecular mass = 194.06
Conclusion: Pass

4 Analysis of the provided digital IR data:

Amount of functional groups given: 3
Amount of functional groups that were part of the routine and accurracy >80%: 2
Output true machine (>80%): 2/2 Pass
Output true owner (>80%): 2/2 Pass
Output false machine (>90%): 0/2 Pass
Output false owner (> 90%): 0/2 Pass

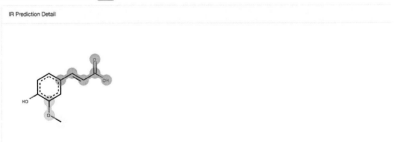

#	SMARTS	Machine Confidence	Machine	Owner
1	c-,:O-,:C	92.53 %	⊘	
2	c-,:O	91.48 %	⊘	
3	c/C=C/C(=O)-,:O	0 %	⑦	

Figure 4-29. Analysis details of MS and IR.

As to the MS analysis details, if the exact molecular weight or the weight of adducts with a proton, sodium- or potassium-cation (according to the most likely adducts that are formed in mass spectrometry) can be identified in the extracted spectrum peaks, the data is labeled as valid (green). Moreover, in the IR analysis detail, it is considered as "pass" when maximum one failure is found for model accuracy >/= 80%, and zero failure is found for model accuracy >/= 90%. Figure 4-29 shows the analysis details of MS and IR.

4.9 Discussion

In this chapter, ChemSpectra, an example of a comprehensive tool for the analysis of spectroscopic data, is presented. It contains an interactive user interface with file parsing and composing of IR, NMR, and MS spectroscopic data. External services can be included in the functions of ChemSpectra: as an example, NMRShiftDB service is connected to offer a computer-assisted method of NMR data curation. Deep learning models are trained to recognize functional groups in IR spectroscopic data automatically. In the end, a conclusion is provided to the user that combines quality control and a quick check of analytical data, giving a measure for quality assurance. This quality of data is given by a number suggesting a confidence level of combined analyses in the Chemotion-ELN.

5 Main Part - Reaction Prediction

A chemical reaction generates new products that have the potential to be used for new applications. The reactivities that lead to a particular mechanism of reaction of certain molecules are not entirely elucidated yet, and novel reactions or reaction mechanisms are constantly found and identified. The fundamental theories are not fully understood, and in consequence, many discoveries are based on coincidence, and a reliable prediction of new synthetic sequences is currently hardly possible. Integrating an A.I. system to explore chemical reactions could reduce human loading and increase the speed of advancement in many fields, e.g., material science and drug discovery.

In this thesis, reaction templates and condition predictions using the template-based approach are investigated. Several publications also adopt the template-based approach to solve problems.[45,50] The reaction template prediction infers possible templates given starting materials, while the condition prediction infers suitable conditions required for a reaction to take place. Therefore, they are a forward-pass transformation from features of starting materials to the output. Forward-pass follows the general convention in chemistry: input starting materials, react under specific conditions, and generate products. On the contrary, the retrosynthetic route planning problem is a backward-pass, since products are input and starting materials are outputs.

The work in this thesis related to this topic is not finished, and, nevertheless, it serves as a preliminary exploration of reaction predictions. Several methods of improving the current methodology are proposed and analyzed.

First, the full template extraction procedure with the restoration mechanism can verify the template correctness. False templates can be avoided to be added to the dataset without mixing into the model training, and besides, it is a vital tool to review and improve modules of a complex process.

Second, the temperature prediction is proposed to demonstrate the possibility to conclude reaction conditions from published data. This prediction could enable an automatic condition recommendation system integrated to the ELN.

In the end, the reaction template prediction is further investigated by adding condition parameters. Existing works related to this topic are conducted only with starting materials as input, not including environment contribution, which is essential in the real-world experiment.

It is worth to mention that two datasets are used in this chapter: the USPTO (the United States Patent and Trademark Office) database and the Reaxys database.[63] It is beneficial to use two datasets due to accessibility and level of detail. Reaction examples of the USPTO database contain patents granted between 1976 and 2013. They are parsed by LOWE *et al.*[65,66] and cleaned and filtered by researchers.[64] This dataset is publicly accessible, but its reaction information is, due to the patent protection purpose, not a detailed disclosure of research. On the contrary, the Reaxys database, a chemistry information database, preserves explicit reaction details since it includes published journals which are research-oriented data. However, the full download of the Reaxys database requires an additional license and agreement.

The template extraction procedure is evaluated on the USPTO database, and other researchers could review this procedure after the project is open-sourced. Predictions are performed on the Reaxys database due to the inevitable demand for detailed reaction conditions as model input parameters. Only small amounts of reactions are used here for fast prototype analyses to gain knowledge of the topic.

5.1 Reaction Template

A template, which is represented by SMARTS descriptions, indicates changing elements in a reaction.[41,99] Reactions can then be categorized into template patterns, and those of the same SMARTS template are assigned to the same class. Given starting materials and a template, products can be deduced from matching components for synthesis product prediction. The question is a multi-class classification problem to predict which template will be the correct one.

Table 5-1 is an example to explain a reaction template. In both reactions, an iodine atom is replaced by a benzene ring, giving the same reaction template for both reactions. Therefore, they are assigned to the same class. Assuming reaction A is in the training set and reaction B is in the test set for the forward-pass prediction problem, during the training phase, starting materials of reaction A are the input, and the target is the label of this exact template. In the ideal scenario, it is expected to predict the same template when inputting the starting materials of reaction B in the inference phase. By template matching, 3-phenylpyridine, C1=CC=C(C=C1)C1=CC=CN=C1, should be the output.

Table 5-1. Two reactions, having the same template, are as assigned to the same class.[a]

Reaction A: ClC1=CC=C(I)C=C1.C1=CC=CC=C1>>ClC1=CC=C(C=C1)C1=CC=CC=C1 Reaxys id = 29417793
Reaction B: IC1=CC=CN=C1.C1=CC=CC=C1>>C1=CC=C(C=C1)C1=CC=CN=C1 Reaxys id = 1533333 
Template: [I;D1;H0:7]- [c;H0;D3;+0:4](:[c:5]):[c:6].[c:1]:[cH;D2;+0:2]:[c:3]>>[c:1]:[c;H0;D3;+0:2](:[c:3])- [c;H0;D3;+0:4](:[c:5]):[c:6] 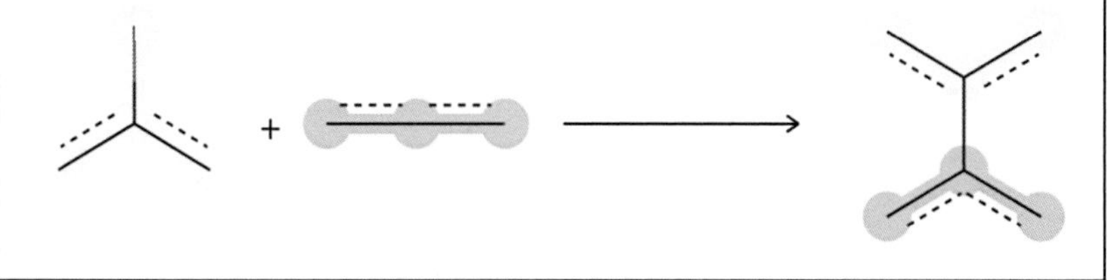

[a] The orange structure in one of the starting materials of the template is part of the ring. Although it is a linear drawing, the dashed line still indicates the ring property.

5.2 Template Extraction

A reaction template contains a reaction core and neighbors defined by radius. The reason to include neighbors into the template is to reduce ambiguous patterns and to define the template more specifically. A suitable radius has to be determined considering two aspects: on the one hand, the more neighbors are included, the more specific the chemical environment is focused. On the other hand, a larger radius increases training difficulty due to a lower number of examples per template. Therefore, a reasonable tradeoff has to be considered when defining neighbors.

There are four essential steps to achieve template extraction. Figure 5-1 illustrates these steps. First, in the data cleaning step, several methods to normalize input data ensure the same input format of reactions from different databases. Second, atom mapping matches the same atoms between starting materials and products. Therefore, atoms can be tracked during a reaction. By extracting atoms that have changed the chemical environment during a reaction, we can extract the reaction core in the third step. In the end, neighbor atoms of the reaction core are included to achieve more specific patterns.

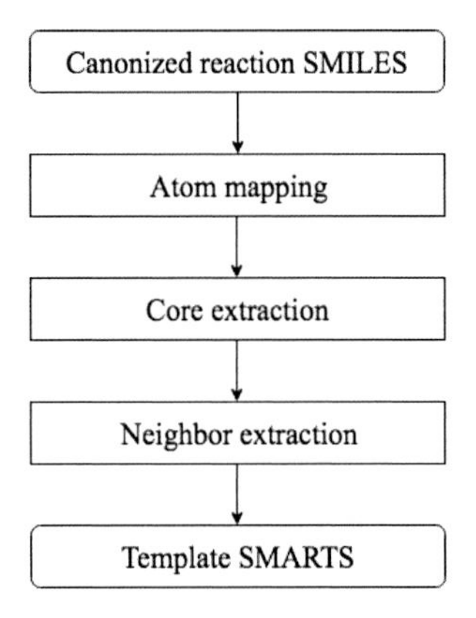

Figure 5-1. Steps of template extraction.

5.2.1 Data Cleaning

In the beginning, several methods are applied as data cleaning, which is similar to the work done by SCHWALLER *et al.*[49] with some modifications to adapt to our models. Reactions are initially recorded in several formats, e.g., SMILES and SDF, from multiple databases. They are converted to SMILES as an input standard in the first place, and then, original atom mapping is removed. Mapping is executed later to synchronize all data. One molecule could have several possible valid SMILES, and in order to overcome this uncertainty, canonization is conducted using RDKit. Role assignments are skipped because we rely on the database assignment. Moreover, there is no tokenization since the fingerprint is used in predictions. In the end, reactions in canonized SMILES are the input to the following modules.

5.2.2 Atom Mapping

An example of atom mapping is illustrated in Figure 5-2 using the previously depicted reaction A (see Table 5-1). The bottom image contains serial numbers designated as mapping tags. The goal is to extract the core, which has minimum noise from unchanged elements.

The reason why we need atom mapping before the core extraction is explained by an example. With visual perception, the human can identify the reaction core with the assistance of mapping tags. Each atom in the starting materials is given a unique number as an identifier, and these identifiers are mapped to the products. From the mapped reaction, the iodine atom with id=14 is removed, and carbon atoms id=5 and id=8 are linked by a single bond. Other atoms are kept without alterations. Therefore, the reaction core with radius 0 has atoms ids: 5, 8 and 14.

In order to obtain a match of the atoms of products and starting materials automatically, cheminformatics tools need to be applied. There are several tools equipped with the atom mapping function: ChemAxon,[100] RDT (Reaction Decoder Tool),[101] Indigo toolkit,[102] ICMap,[103] and ChemDraw.[80] In this thesis, RDT is adopted as the primary mapping tool since it is an open-sourced software and has state-of-the-art performance.[104]

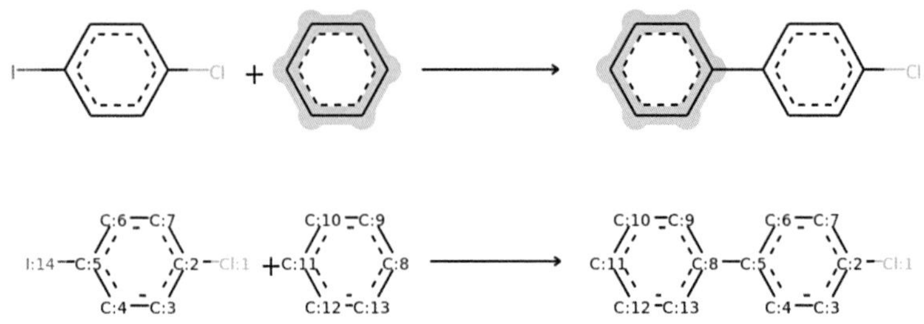

Figure 5-2. Atom mapping to identify changes.

5.2.3 Core Extraction

A core extraction tool generates the SMARTS of the template pattern by systematically comparing the chemical environment of atoms and bonds one-by-one. In a recent publication from COLEY *et al.*,[41] the authors managed this challenge and provided the source code, which is adopted as the basis in this thesis.

According to COLEY *et al.*, the core and template extraction algorithm is developed based on RDKit. Product atoms and starting material atoms are looped to find the same atoms from both sides. If mapping tags are the same for one product atom and one starting material atom, they are the same atom, and properties listed in Table 5-2 are analyzed to know whether the chemical environment is modified. If it is modified, the reaction core includes the atom. Different atom mapping tags from two sides are different atoms, and further comparison is skipped.

Table 5-2. If an atom belongs to the reaction core, at least one of the following factors is altered during the reaction.

Reaction core factors
- atomic SMARTS
- atomic number
- total number of hydrogens connected
- formal charge
- number of radical electrons
- aromatic
- bond angles
- connected bonds
- chirality

5.2.4 Neighbor Extraction

The extension of the reaction core to neighbors leads to the selection of a more specific chemical environment, allowing to narrow down the scope during a similarity-based search. If a reaction core radius is 0, the template includes no neighbors. With a radius of 1, first neighbors of the core are included, while radius 2 extents the template further to the second neighbors. Examples are depicted in Figure 5-3. Therefore, in the case of radius 1, the atom ids: 4, 5, 6, 8, 9, 13, and 14 are highlighted. Atom ids: 3, 4, 5, 6, 7, 8, 9, 10, 12, 13, and 14 are marked in the case of radius 2. In case of a larger radius it takes more atoms as the template.

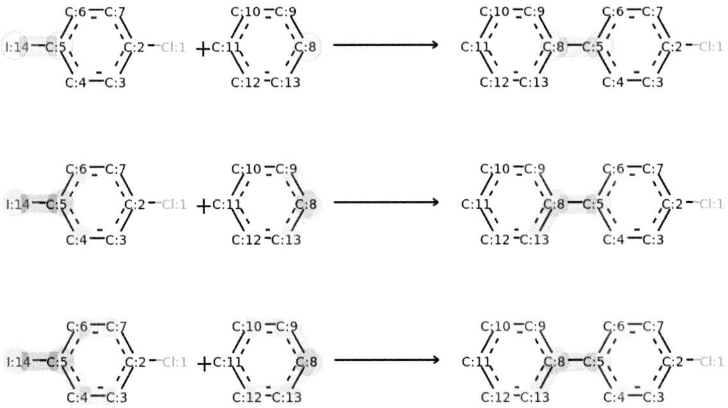

Figure 5-3. Atom changes with different radius. From top to bottom are radius 0, 1, 2. Pink areas highlight the changes in starting materials, and green parts are those of the product.

5.3 Template Verification - Restoration

The template extraction process is in an open loop form, where there is no feedback mechanism to ensure the correctness of the result. Although the extraction is designed based on chemical logics in Table 5-2, there is no guarantee that the process is complete to cover all reactions. This process is adopted to extract building blocks, reaction templates, to categorize reactions in many works, but a thorough analysis is absent. A systematically and automatically algorithm is desired to verify the outcome effectively, which is the first improvement provided in this section.

In this thesis, a verification method, restoration, is proposed, which is an algorithm to keep correct templates and avoid building models on false templates. Incorrect templates mixing into datasets would lead to inaccurate predictions. Figure 5-4 illustrates restoration integration with the template extraction process. When the verification is correct, the output is the template SMARTS. Conversely, none is returned, recorded as null in the database, for the fail verification. False cases will be examples to improve the template extraction procedure. Therefore, the second advantage of restoration is to enable a reliable way to find a defect and blindside in the complex extraction process.

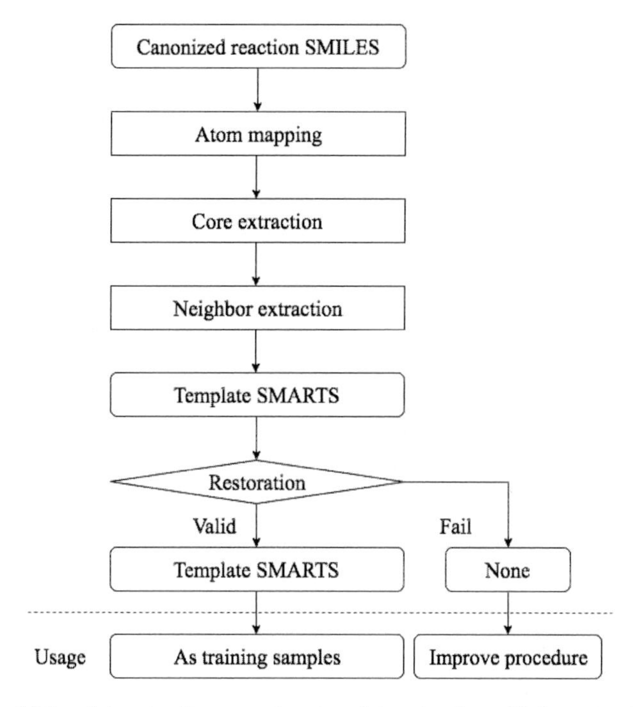

Figure 5-4. The full template extraction procedure: template extraction with the restoration.

The restoration method combines starting materials and the extracted template to render deduced products. The verification is done by comparing deduced products and original given products. If they match, the template is valid because there is no information loss after transformation. For the reaction prediction problem, when given starting materials and the predicted template cannot generate the correct products, the whole extraction process is still useless because no product can be inferred and not enough information is kept. The restoration is built under this concept to verify extraction. In the publication of PLEHIERS *et al.*,[99] the authors adopt a similar approach to verify the extracted templates.

The full template extraction procedure, including the restoration, is applied to the first 2000 reactions of the USPTO database. A third party can use this open-access dataset to reproduce the procedure after the project is open-sourced, which is the reason why we use this database. The full dataset analysis can run without additional legal workflow. Despite the patent-oriented nature, the database is used to demonstrate the ability of the extraction procedure, and information distortion because the patent protection purpose can be neglected.

Two mapping tools were used to demonstrate the developed procedure: RDT and ChemAxon. The template extraction using RDT can yield higher accuracy than that of ChemAxon as shown in Table 5-3. False cases are not the same in both mapping tools, because different algorithms exhibit different performances in different reaction types.[104] For example, in a study of metabolic reactions, RDT achieves the highest mapping accuracy in reactions catalyzed by oxidoreductases and ligases, but other tools outperform RDT in reactions catalyzed by transferases and isomerases. Moreover, a hybridized method using RDT, ChemAxon, and other tools, could be a possible way to achieve higher accuracy since this integration can map correctly on a broader range of reaction types.

Table 5-3. Apply the full template extraction process to 2000 reactions from the USPTO database.

Mapping Tool	**RDT, Reaction Decoder Tool**	**ChemAxon**
Valid count	1847	1754
Valid rate	92.3%	87.7%

Although these reactions are only a small part of the whole database, this preliminary step is a good starting point to demonstrate that the procedure can gauge different mapping tools, potentially other modules in the procedure, systematically by accuracy numbers. Before, related works have no measurement of the template extraction quality or merely selected a few examples to justify their points.[41,45,50] A review of outcomes is done manually without a robust automatically operation. Analyzing results based on numbers is a scientific way to make a decision.

Once the accuracy is known, tools and algorithms can be selected to be applied to the model. Reactions without valid templates could be reviewed to improve the process later. Moreover, this procedure could also be integrated into the ELN, and assist users to document reactions with reasonable role assignment which could be included into the machine learning afterwards to minimize information loss due to inaccurate assignments.

During the processing of reactions according to the designed method, low throughput is encountered, i.e., a long mapping time for all reactions. In order to solve this, reactions are sent to BwUniCluster servers[105] for parallel mapping. The final templates are sent back and stored in the source database. Figure 5-5 and Table 5-4 show the architecture and performance measurement. The measurement is conducted with only 100 reactions from the USPTO database.

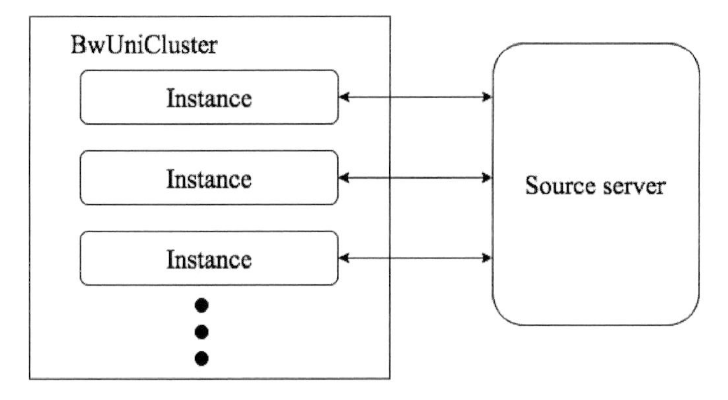

Figure 5-5. The distributed computing architecture.

Performance improvement is achieved by distributed computing. Processing of the full extraction by running 4 instances on one Mac air notebook results in a reduction of ~30% of the time, because CPU usage increases. Running 8 instances on BwUniCluster can further reduce time to ~20% of the initially spent time. Nonlinear improvement with respect to instance count results from the fact that some reactions take a longer time for mapping and each instance has different capacity between the server and the laptop. For efficient usage, it is desirable to have a resource management mechanism to allocate computing resources and distribute reactions uniformly. In massive extraction, around 40 instances could be initialized, which has already been verified, on BwUniCluster based on real-time traffic and resource.

The reaction template is one powerful tool to understand chemical reactions. This fact is the driving force to propose distributed computing for the procedure in this thesis. The USPTO and the Reaxys databases contain more than 1 million reactions, and as time goes by, published reactions will continue to accumulate. Meanwhile, new methods to improve reaction structure, e.g., role assignment, will lead to a possibility to redo the work again. Therefore, increasing the speed of the full procedure is imperative. By reducing time on template extractions, newer methods or creative applications, where templates are fundamental elements, can be tested efficiently.

Table 5-4. Performance measurement of the full process time by distributed computing.

Configurations	Time
Mac air 2015 (1 instance, one 4-core CPU)	426 sec
Mac air 2015 (4 instances, one 4-core CPU)	315 sec
BwUniCluster (8 instance; each has 4-core resource)	80 sec

5.4 Predictions

There are two predictions in this section: reaction temperature and template.

5.4.1 Temperature Prediction

In this part, the required temperature for a reaction is predicted given the starting materials. In many cases, not only starting materials but also conditions define products if there are several options. Building a model that infers the required temperature for a reaction could help chemists to find the optimal conditions when planning the reaction. This prediction is a possibility to estimate an adequate reaction temperature with help of previous research data. Six thousand reactions are picked from the Reaxys database[63] because this commercial database preserves temperature conditions from published reactions. Reaction templates are extracted based on the full verification method, as mentioned in the previous section.

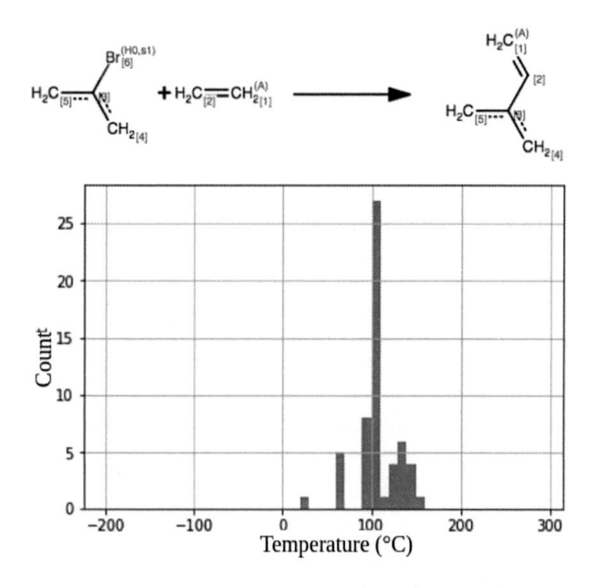

Figure 5-6. Temperature distribution of a reaction class.

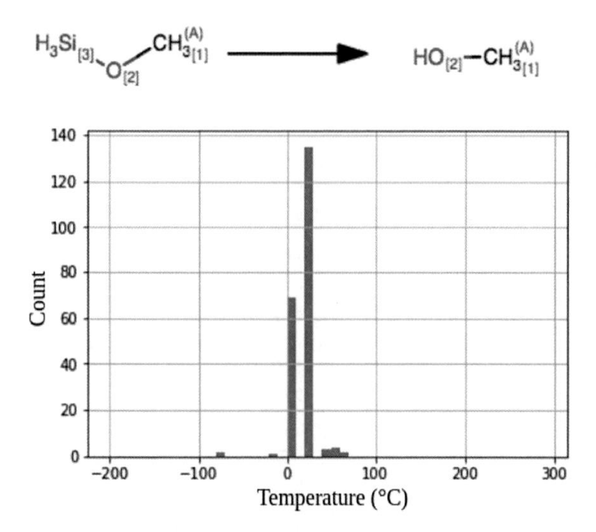

Figure 5-7. Temperature distribution of a reaction class.

First, statistics of reaction template versus temperature are analyzed. Here, two reaction templates are selected as examples. They are illustrated in Figure 5-6, and Figure 5-7. The x-axis represents the temperature for reactions parsed from the database, and the y-axis represents the appearance count of a temperature range. For example, (20, 30] means that reactions are counted when the highest temperature is equal to or larger than 20 °C and smaller than 30 °C. Their count is displayed as a bar at x=20. The majority of the reaction class in Figure 5-6 is around 90 - 150 °C, and that in Figure 5-7 is around 0 - 30 °C.

The data denotes a correlation between template and temperature. A deep learning model with four layers is built to predict the required temperature by inputting starting materials. This architecture is similar to the work for the reaction template prediction.[45,50] Input features are fingerprints of starting materials in the ECFP4 form with 1024 bits. Since there could be one or more starting material, each fingerprint bit denotes the existence of a specific fragment in the starting material. The output is the degree in Celsius. This regression problem has the MAE (mean absolute error) loss function. Dataset split ratio is 7:2:1 for training, validation, and testing sets. The model architecture is illustrated in Figure 5-8. Considering +/-10 °C deviation is correct, and the accuracy is 73.05%. Figure 5-9 shows the loss curve.

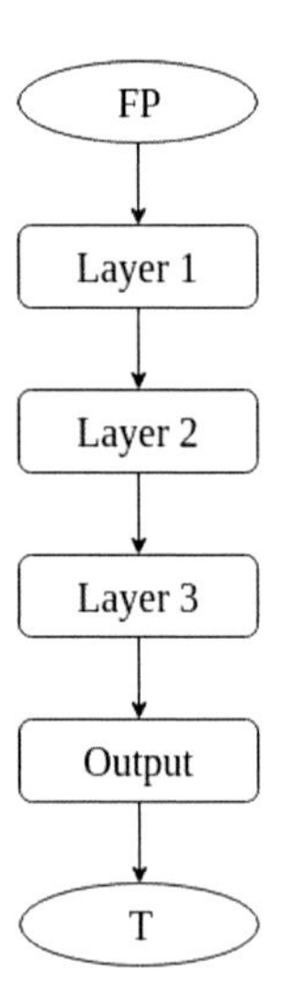

Figure 5-8. The architecture of temperature prediction.

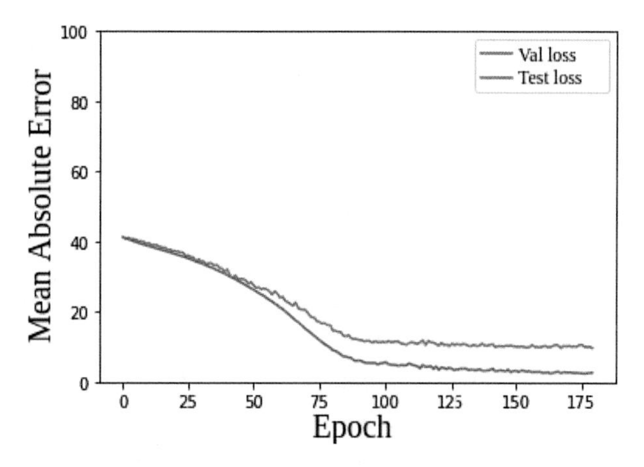

Figure 5-9. Validation and testing loss.

5.4.2 Template Prediction

Several publications[41,45,50] adopt the template prediction. However, the effect of including additional features, like reaction temperature, to improve accuracy is still unknown. Reaction conditions are essential factors that dominate the reaction mechanism and can affect the formation of products. There are 2961 reactions selected for this section, and, in total, 26 template classes are extracted. Each class has more than 50 reaction examples in order to learn from reasonable sample amount. These classes contain temperature affected reactions in order to demonstrate the idea.

Several models are built to analysis whether adding temperature can improve the accuracy of the template prediction. These models are divided into two groups: the experimental group and the control group. The experimental group merges fingerprints and temperature parameters as input features, while the control group only contains fingerprints. Therefore, if the experimental group can demonstrate accuracy improvement compared to the control group, the assumption that temperature is one key factor for the template prediction is valid.

Basic input features are fingerprints of starting materials in the ECFP4 form with 1024 bits, the same as that in the previous section. This fingerprint model is adopted in published research,[45,50] One-hot encoding is applied to merge temperature into binary features, and a six-bit representation indicates that reaction temperature is in a specific range. Only the highest temperature during reaction is considered, and it is possible to add the lowest temperature into bits for further development. The reason to adopt one-hot encoding is that it is simple and widely used in the deep learning study. Although temperature can be inputted into the network directly without additional encoding, a proper normalization is needed to avoid the numerical stability problem.

The temperature range is defined based on the following facts, and the additional margin is considered to include different documentation conventions. Room temperature is15 - 30 °C in the fourth bit and this can consist of different possible cases like 18, 20 or 25 °C. The last bit is decided based on the boiling point of water, 100 °C, and temperature larger than 90 °C is considered with a 10 °C margin. Zero degrees is the water freezing point, and it separates second and third bits. This definition is not perfect, but this is the first step to approach the problem.

Figure 5-10 is an example where the highest temperature of the reaction is room temperature. The output layer is 26 bits as a multi-class classification problem, and there is only one positive output bit for an example.

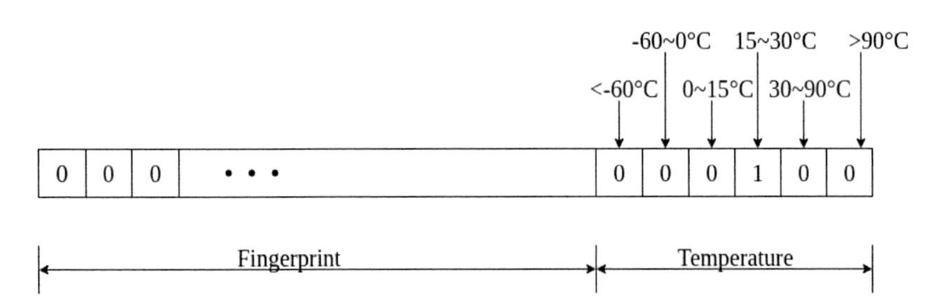

Figure 5-10. Input features for the experimental group. In this case, the highest temperature is room temperature.

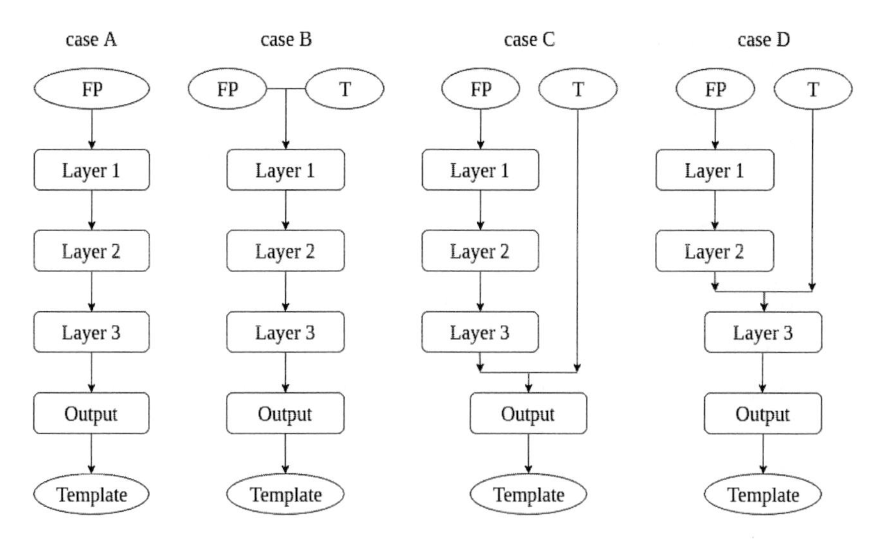

Figure 5-11. Deep learning models adopted in this section.

Table 5-5. Accuracies of the control group and experimental groups.

Case	**A**	**B**	**C**	**D**
Accuracy	77.04%	77.52%	78.42%	77.02%

Four layers, including one output layer, multi-layer perceptron neural networks are constructed, and they are illustrated in Figure 5-11. Case A is the control group, and cases B, C, and D belong to the experimental group with temperature bits as parts of input features. Cases A and B have the same model architecture with different input features: case A contains only fingerprints of starting materials while case B contains both fingerprints of starting materials and one-hot encoding of the highest temperature. Cases C and D are variants to analysis the effect of inputting temperature bits in different layers. Table 5-5 lists accuracies, and there is no substantial proof that adding temperature bits can boost accuracy. Although there is no profound improvement, including other conditions and different possible improvements will be further studied in the future.

5.5 Discussion

Currently, the dataset contains only a subset of the database. Although temperature-related reactions are included, some types of reactions, e.g., combustion and metabolism, which have higher temperature dependency, should be analyzed closely. Narrowing input dataset to these high-temperature-dependent reactions could increase temperature weighting.

Data curation and preparation is a critical stage since it is the very first step of the input. Here, database providers define roles. However, a robust and automatic role assignment is desired. SCHNEIDER *et al.*[106] propose an algorithm to assign roles effectively by fingerprints, and afterward, this will be implemented into the full procedure.

Different models and temperature encoding can be tested. The language translation model and the graph model have shown higher accuracy in the USPTO dataset compared to the MLP model with fingerprint input features adopted in this thesis. However, the language translation model and the graph model are non-template based. The message of higher performance delivers a hint that it is worthwhile to include them into consideration. A fusion of the LSTM network and condition parameters could be one way to embed different features. More bits to represent specific temperature region and appending more trainable layers at temperature input should also be analyzed. Nevertheless, the prototype is present in this thesis, where variants could be constructed by adding or modifying submodules.

6 Main Part - Reporting

Exporting data from the report module of the Chemotion-ELN with predefined formats is essential. This module supports scientists in their efforts to share and submit data to others. When design, procedure, and details are well organized and formatted, creating and reading a report is a pleasant enjoyment with less ambiguity.

The reporting module is an indispensable part of modern ELNs. In this thesis, not only essential integration with the Chemotion-ELN is achieved, but also unique features are implemented. This module provides several reporting formats. Supporting information is designed to be directly publishable. One article published with this feature is already available.[107] Structured data readable by the machine, which is the co-working result with one of the leading publishers, enables seamless data transferring to other institutes. Achieving FAIR data principles, findability, accessibility, interoperability, and reusability are the central vision of this module.[78]

Moreover, chemists can access molecular structures via ChemDraw bidirectional data binding, which is a crucial standard in Chemistry. A ruby gem is created to import RInChIKeys[108] into the module. These key characteristics are introduced in this chapter.

The generation of reports using the information collected in the ELN, which is requested frequently, is helpful for scientists. Besides, an efficient reporting function is of high importance with respect to the support of a digital exchange in chemistry. The main advantages can be summarized as follows:

(1) An efficient generation of the report can save time for users. What users need to do is to input their daily activities regularly, and then, reports can be created by selecting items and clicking a few options within a short time. Reorganizing scattered information after some time is avoided.
(2) Generated reports are well-formatted and can be understood easily in comparison to manually written documents due to the predefined structure.
(3) A summary of the stored information minimizes the risk of error during the preparation of a reaction.

(4) A reporting function allows adding machine-readable data in a straightforward manner which allows meeting the FAIR data principles without any additional effort.

During the implementation of the reporting function, different aspects have to be considered. The two most important ones are:

(1) Minimizing the effort of the user (user input) is highly desirable while in parallel, allowing high flexibility adapting to the diverse needs for different styles and habits. This module has to be designed in a way that very little additional work has to be added once created.
(2) The implementation of existing standards for chemistry research is decisive to ensure the acceptance of the generated reports by the community. Therefore, common standards in chemistry have to be investigated, and appropriate conventions have to be embedded.

Table 6-1 Report types

Type
Standard
Supporting Information
Supporting Information – Spectra
Supporting Information - Reaction List (.xlsx)
Supporting Information - Reaction List (.csv)
Supporting Information - Reaction List (.html)

6.1 Report Types

As listed in Table 6-1 , there are six report types for different scenarios. These report types can be split in two main groups: (1) standard report and (2) supporting information.

6.1.1 Standard Report

The standard report is for internal group reviewing or information exchange. Its format can be further customized by editing the document template. A standard report may contain the two main sections, samples, and reactions that are enriched with further information according to the settings defined by the user.

6.1.1.1 Sample Section

Figure 6-1 shows an example of a sample section. The report allows the summary of all necessary information given for the sample consisting of automatically processed information and data entered by the user.

On the title area, several primary identifiers for the target are listed: IUPAC name, short label, and collection label. The IUPAC name provides a systematic way to interpret the molecular structure. The collection title suggests the correlation to the project, while the short label can be used as the search key in the system where less than eight digits should be typed in.

The sample structure is illustrated as a two-dimensional image, which increases reading efficiency. By clicking on it, the structure information is rendered in the ChemDraw application, in which further editing is reflected to the image after switching it off. If selected, the results from analytical measurements, like ^{1}H and ^{13}C NMR, IR, and MS spectroscopy, are added in the form of peak arrays. The prerequisite is that the user of the ELN gives this information in the analysis section.

Analyses are pre-encoded as description templates, and experimental data can be inserted into templates easily. There is no need to select complex symbols and to adjust formats manually. These parameters are key factors to reproduce or trouble-shoot the measurement in the future.

Date: 17.06.2019

6-(1,3-dithiolan-2-ylidene)octan-3-one (SSS-162)

Collections: : Final Order, My project with Yu-Chieh Huang, My project with Yu-Chieh Huang

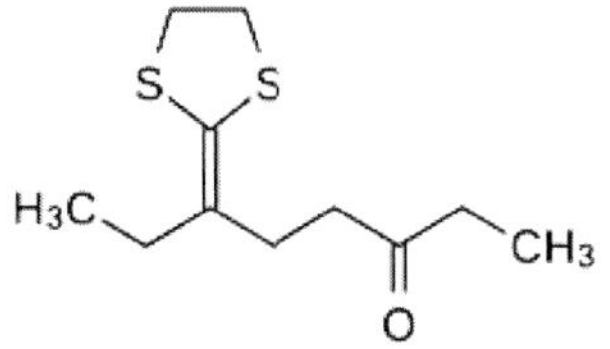

^{1}H NMR (400 MHz, $CDCl_3$, ppm), δ = 1.01 (t, *J* = 7.5 Hz, 3 H), 1.06 (t, *J* = 7.3 Hz, 3 H), 2.16 (q, *J* = 7.5 Hz, 9 H), 2.40 - 2.47 (m, 4 H), 2.49 - 2.56 (m, 2 H), 3.33 (s, 4 H). ^{13}C NMR (100 MHz, $CDCl_3$, ppm), δ = 7.8, 12.0, 29.8, 30.2, 35.8, 37.6, 37.7, 40.0, 127.5, 129.3, 211.1. EI (m/z, 70 eV, 20 °C): 230.2 (65), 181 (31), 159 (100). HRMS ($C_{11}H_{18}OS_2$): calc. 230.0799, found 230.0800. IR (ATR, ṽ) = 2963, 2928, 1710, 1600, 1457, 1419, 1359, 1278, 1147, 1111, 1045, 987, 913, 848, 827, 685 cm^{-1}.

Figure 6-1. A sample section of the standard report.

6.1.1.2 Reaction Section

In accordance with the preferences of the users, this section is very similar to the reporting of samples. The user defines the amount and type of data for most of the given information in the report.

In the title area, short label and collection title have the same function as those in the sample section, as the search key and project identification are given in any case, without user selection. Moreover, a reaction status is denoted, for which stage it is when composing the report. Some reactions and analyses may take several days, and the status is needed for others to track the schedule. There are six stages: planned, running, successful, not successful, done, and analysis pending.

All other information can be selected but is not mandatory in the standard model. The options for the data to be selected according to the user's preferences are given in Figure 6-2. An example of a reaction section is shown in Figure 6-3.

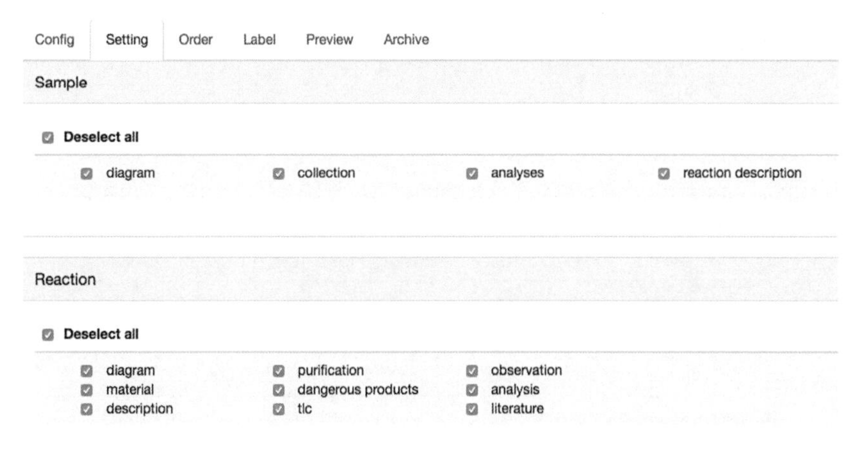

Figure 6-2. Optional information in the standard model can be selected in the setting panel.

Typical information for reactions (not relevant for the reporting of the samples) is the reaction scheme which typically contains a drawing of starting materials, products, reactants, solvents, and temperature. Readers can understand these roles quickly based on chemical symbol conventions: the arrow is in the central part, starting materials are drawn on the left side of the arrow and products are placed on the right side of it. The arrow-head pointing to the right side indicates the reaction direction. Reactants are placed on the top of the arrow, while solvents and temperatures are put on the bottom of it. This image, which has the same function as that in the sample section, can link to the ChemDraw application directly by double-clicking on the image.

A material table, after the reaction scheme, describes samples' amount by giving mass and volume. Equivalent and yield represent relative quantity to a reference. Samples' IUPAC names (if available; if not – the sum-formula) and short labels at the beginning of each row are identifiers.

Date: 17.06.2019

SSS-R50 Reaction: According to General Procedure 2a Successful

Collections: Final Order, My project with Yu-Chieh Huang, My project with Yu-Chieh Huang

Acetonitrile

55%

Formula	Mol mass	Mass [g]	Volume [mL]	Density [g/mL]	mmol	Equiv/yield
C6H11BF4S2 (SSS-27-10)						
S C6H11BF4S2	234	**0.250**	0.250	1.00	1.07	1.00
C5H8O (reactant)						
R C5H8O	84.1	**0.108**	0.128	0.844	1.28	1.20
C11H18OS2 (SSS-162)						
P C11H18OS2	230	**0.135**	0.135	1.00	0.585	55%

Solvent(s): Acetonitrile (5.00ml)

Description:

Type of Purification: Flash-Chromatography

Dangerous Products: Flammable liquid (Class 3), Toxic and very toxic (Class 6.1)

TLC control: Rf-value: 0.58 (Solvent: cyclohexane/ethyl acetate 20:1 + 1% NEt_3)

Observation:

The obtained crude product was purified via flash-chromatography on silica gel using cyclohexane/ethyl acetate 10:1

Analysis:

C11H18OS2 (1H NMR)

^{1}H NMR (400 MHz, $CDCl_3$, ppm), δ = 1.01 (t, J = 7.5 Hz, 3 H), 1.06 (t, J = 7.3 Hz, 3 H), 2.16 (q, J = 7.5 Hz, 9 H), 2.40 - 2.47 (m, 4 H), 2.49 - 2.56 (m, 2 H), 3.33 (s, 4 H).

C11H18OS2 ()

^{13}C NMR (100 MHz, $CDCl_3$, ppm), δ = 7.8, 12.0, 29.8, 30.2, 35.8, 37.6, 37.7, 40.0, 127.5, 129.3, 211.1.

C11H18OS2 (Mass)

Generated by Yu-Chieh Huang page 1 of 3

Figure 6-3. A reaction section in the standard model.

Further information is appended after the material table, these are solvents, descriptions, purifications, dangerous products, TLC control, observations, analyses, and literature. Each sample or reaction occupies a page or more, and it always starts from a new page for a comfortable visual perception.

6.1.2 Supporting Information

The supporting information delivers data as supplementary to journals, in which readers can grasp the full pictures of synthesized products. In order to maintain the same standard across individual articles and journals, the format of the supporting information should be well-defined and only customized concerning very few circumstances.

The standardization is crucial to compare results, to foster a fast understanding of the given information, and to be able to be read by machines in the future. Currently, the prerequisite for sustainable research is not fulfilled by most of the publications, and the herein described procedure to gain standardized supporting information may serve as an example of how to solve this challenge later. In order to meet the requirements of all journals in a comprehensive way, five types of supporting information have been designed, each giving a particular type of information that is commonly requested by the publishers (please see the list in Table 6-1 with the same prefix: Supporting Information).

Experimental Part:

1 Versions

Version InChI (1.04), Version SMILES (Daylight)

2 General remarks

3 General procedures

3.1 General Procedure 2a (SSS-R55)

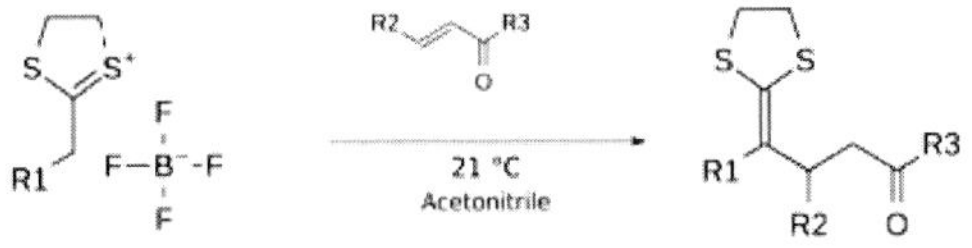

The dithi(ol)anylium tetrafluoroborate {1} (1.00 equiv.) was dissolved in dry acetonitrile {3} in a glass vial at room temperature if not otherwise stated. The α,β-unsaturated ketone {2} (1.20 equiv.) was added in one portion, the reaction was stirred at room temperature if not otherwise stated and was observed via TLC control. To all reactions, silica gel (3 g) was added after 1 h of reaction time and the solvent was removed via evaporation under reduced pressure. Even though some of the reactions were observed to be finished faster, all of them were reacted for 1 h to allow a good comparison of the results.

4 Synthesis

4.1 6-(1,3-dithiolan-2-ylidene)octan-3-one (**xx**)

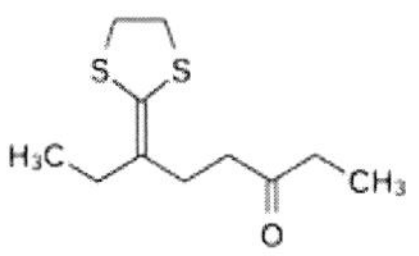

Name {P1|**xx**}: 6-(1,3-dithiolan-2-ylidene)octan-3-one; Formula: $C_{11}H_{18}OS_2$; CAS: - ;
Molecular Mass: 230.3900; Exact Mass: 230.0799; EA: C, 57.35; H, 7.87; O, 6.94; S, 27.84.
Smiles: CCC(=C1SCCS1)CCC(=O)CC
InChIKey: YDJFXLVHPNACDY-UHFFFAOYSA-N

According to General Procedure 2a: {A|**xx**} C6H11BF4S2 (0.250 g, 1.07 mmol, 1.00 equiv); {B|**xx**} C5H8O (0.108 g, 1.28 mmol, 1.20 equiv); {S1} acetonitrile (5.0 mL); Yield {P1|**xx**} = 55% (0.135 g, 0.585 mmol).
The obtained crude product was purified via flash-chromatography on silica gel using cyclohexane/ethyl acetate 10:1. R_f = 0.58 (cyclohexane/ethyl acetate 20:1 + 1% NEt_3).

Figure 6-4. An example of the primary supporting information.

6.1.2.1 Primary Part of the Supporting Information

The primary part describes details of the reactions which are conducted for the given publication. In chemistry, most of the experimental descriptions and analytical results are given in the supplemental documents because of the space limitation in articles. The style of the Supporting information is similar to the standard report with the format designed based on current articles from several leading journals in synthetic chemistry. An example of the primary supporting information is shown in Figure 6-4.

There are four parts to the primary supporting information:
(1) Information on the given data and its generation
(2) General remarks
(3) General descriptions
(4) Synthesis part

First of all, InChI and SMILES versions adopted in the document are listed in order to manage across different versions of tools: it is InChI 1.04 and daylight SMILES used in the Chemotion-ELN in September 2019. The general remarks part keeps the user-defined message, which is left as blank for the author to fill in.

The third part is the general procedure. It is a generalized model, from which several reactions can be derived. A reaction image with ChemDraw binding is added, and the description is appended to indicate conditions and steps.

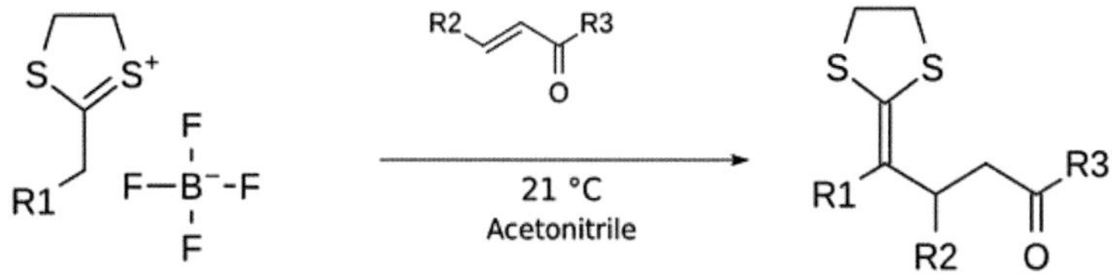

Figure 6-5. A scheme of a general procedure.

This part contains a scheme that represents the given transformation generically, using different residues for side chains to be kept flexible. Distinct reactions for the general procedure have to be assigned by the user of the ELN (for assigned reactions see the following section). An example of such a template, including different placeholders for side chains is given in Figure 6-5. By template matching, the synthesis product has ethyl in R1 and R3. While R2 is a hydrogen-atom. Then, the starting material can be deduced by replacing ethyl at R1 and R3. Figure 6-6 illustrates the product.

Figure 6-6. Template replacement of Figure 6-5: ethyl in R1 and R3. R2 is a hydrogen-atom.

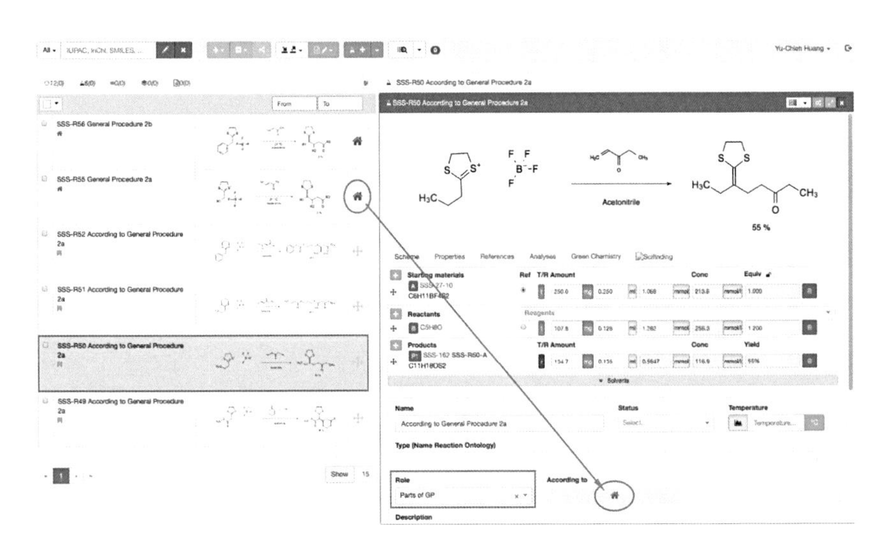

Figure 6-7. A relation among a general procedure and a synthesis is built by selecting 'role' and drag-and-drop the general procedure to the 'according to' field in the browser.

To link the relation between general procedure and synthesis, the user needs to assign them in the user interface: the role field in the reaction scheme page. There are three roles: general procedure, parts of the general procedure, and single. A single reaction denotes a unique case which does not belong to any general procedure. After selecting the role "Parts of GP", a house icon pops out, and the user has to drag-and-drop the general procedure from the reaction list to define which template it is according to, as illustrated in Figure 6-7.

The fourth part, synthesis, includes all products of the article. The user can assign products to the corresponding general procedure according to the commonly accepted process for the generation of supporting information described above.

The product detail is listed, including IUPAC name, formula, CAS number, molecular mass, exact mass, elemental composition, SMILES, and InChIKey. Synthesis information, including description, material table, TLC, product analyses (for details see sample reporting) and literature are appended after it. Here, the material table is the amount of all material. Analyses are peak locations of spectra.

6.1.2.2 Spectra

While in former times, the supporting information mostly contained only the necessary information in written form, more and more journals insist on the submission of a spectra part that includes images of the most relevant analytical data. In organic chemistry, this is, in general, the ^{1}H and ^{13}C NMR spectra or analytical data like HPLC analysis, IR or MS spectra.

The accessibility of spectra is beneficial for the reproducibility of experiments as images are more intuitive for readers to grasp the concept. Moreover, images are also an essential material of the reviewing process as impurities can be detected. Images are more difficult to imitate than a peak list as a result of the analysis. Therefore, the spectra part is the control mechanism in the reviewing process, and it contributes to the curation of scientific reporting.

A prerequisite for the generation is the availability of image files for each analysis to be reported. They can be added from manual scans or digitally saved files. An integration solution of the ChemSpectra (see chapter 4 of this thesis) generates images in parallel to the processing of JCAMP-DX data. Desired files are collected based on the user's selections. In the document, only image formats, i.e., png, jpg, jpeg, are inserted, and other file formats like jcamp, or raw files are discarded. Figure 6-8 shows an example of the spectra supporting information.

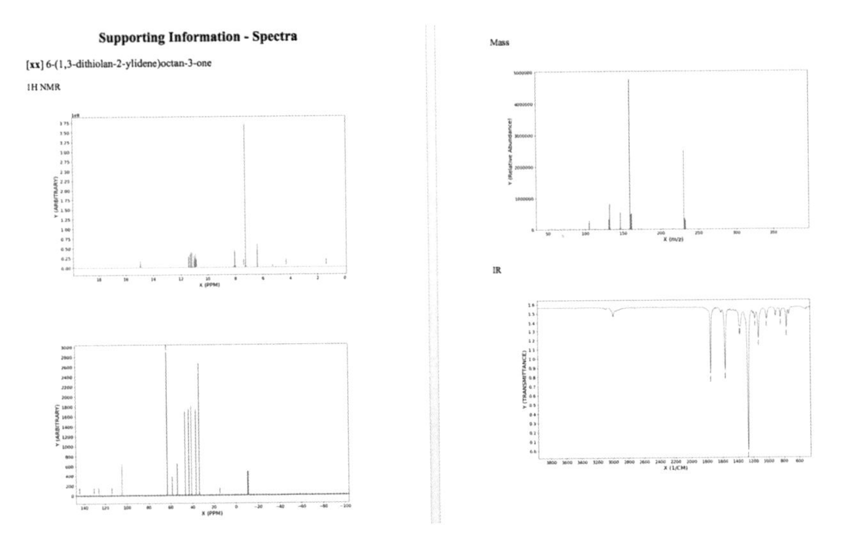

Figure 6-8. An example of the spectra supporting information.

6.1.2.3 Reaction List

Reports are not only helpful for humans, but also for organizations to share and disclose data in a more efficient way. Traditionally, articles were printed and distributed to the readers. Therefore, old articles can be found on the internet with only scanned versions, where the resolution is not always satisfying. Besides, selecting or searching of the full text is not available, since all information is stored as an image and not as machine-readable data.

Furthermore, today most of the publications, including the supporting information, are available as pdf documents. This convention still hinders the search and the reuse of the information as in particular the chemical structures cannot be identified automatically with high precision. To date, there are few optical character recognition (OCR) tools, in chemistry called optical structure recognition (OSR), useful to identify information in schemes and to convert it into text. One obstacle is that the accuracy is in general quite low and OCR causes much misleading information due to a wrong translation of scanned structures.

A full electronic article can overcome those obstacles and can deliver the presentation of chemical structures and reactions. A more concise version of data is also preferred for the machine to read.

To develop a suitable approach to introduce a draft for machine-readable information along with an article to be submitted to a journal, the Beilstein Institute (Frankfurt) joins as a partner. The Beilstein Institute is the publisher of two Platinum Open Access Journals, one of them is the Beilstein Journal of Organic Chemistry. Aiming for complete openness in chemical research, documentation and publishing, the "Beilstein Institute" zur Förderung der Chemischen Wissenschaften[109] is interested in publishing research results in a machine-readable manner to allow standard search engines to keep a record of published items. Such a procedure could facilitate the search and reuse of information via searching published procedures via google.

In order to reach this aim, tables containing samples of reactions were created as a reaction list and are made available in three formats: XLSX, CSV, and HTML. Organizations, like publishers or database providers, can read these files programmatically to their databases. Furthermore, the HTML file can be directly embedded into a website. This end-to-end data transfer integrates and distributes the information seamlessly.

Samples in a table have different fields: label, structure image, InChI, InChIKey, Long-RInChIKey, Web-RInChIKey, and Short-RInChIKey, which are suggested by Beilstein Institut. Among them, Open Babel provides InChI and InChIKey, while the RInChI program[108] creates Long-RInChIKey, Web-RInChIKey and Short-RInChIKey. Figure 6-9 and Figure 6-10 show reaction lists in XLSX and HTML formats.

	A	B	C	D	E	F	G	H
1	Article Reference					Article DOI		
2	reference					doi.		
3	Label	Image	Name	InChI	InChIKey	Long-RInChIKey	Web-RInChIKey	Short-RInChIKey
4	xx			InChI=1S/C18H26O2S	HAPMDIPOMSAKCB-U	Long-RInChIKey=SA-F	Web-RInChIKey=WOH	Short-RInChIKey=SA-F
5	xx			InChI=1S/C13H24O2S	YHZSWQASJCLPOE-U	Long-RInChIKey=SA-B	Web-RInChIKey=YGU	Short-RInChIKey=SA-B
6	xx		6-(1,3-dithiolan-2-ylider	InChI=1S/C11H18OS2/	YDJFXLVHPNACDY-U	Long-RInChIKey=SA-B	Web-RInChIKey=LEIN	Short-RInChIKey=SA-B
7	xx			InChI=1S/C12H16O3S	XOKFBPMVJSDUJQ-N	Long-RInChIKey=SA-B	Web-RInChIKey=VGX	Short-RInChIKey=SA-B

Figure 6-9. A reaction list in the XLSX format.

Article Reference	Article DOI
reference	doi

Label	Image	Name	InChI	InChIKey	Long-RInChIKey	Web-RInChIKey	Short-RInChIKey
xx	[structure]		[illegible]	[illegible]	[illegible]	[illegible]	[illegible]
xx	[structure]		[illegible]	[illegible]	[illegible]	[illegible]	[illegible]
xx	[structure]	[illegible]	[illegible]	[illegible]	[illegible]	[illegible]	[illegible]
xx	[structure]		[illegible]	[illegible]	[illegible]	[illegible]	[illegible]

Figure 6-10. A reaction list in the HTML format.

The developed procedure was implemented into the ELN as one of the available reporting models and was used for the export of information for the publication "Addition of dithi(ol)anylium tetrafluoroborates to α,β-unsaturated ketones".[107] The identifier's list is given on the website of the publisher, allowing to search and retrieve the information on published items via a Google InChIKey and RInChIKey search.

6.2 User Interface

An image of the user interface of the report function is shown in Figure 6-11, and the flow chart in Figure 6-12 explains the required steps to create a report.

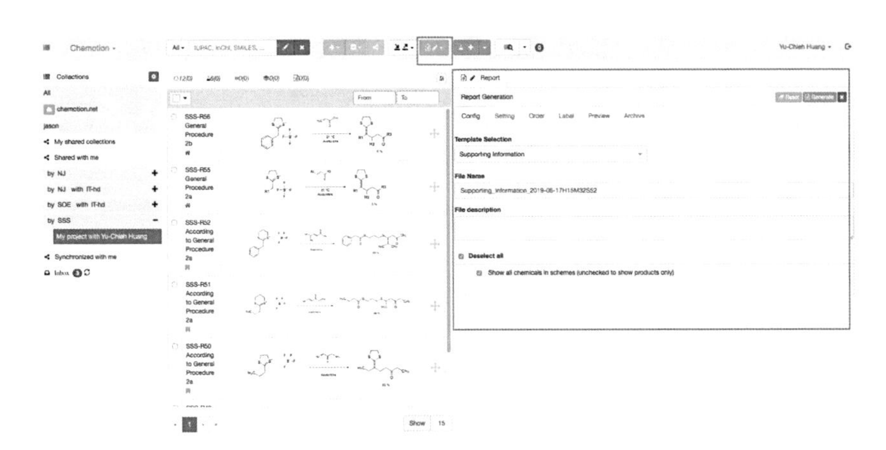

Figure 6-11. Accessing the user interface of the report panel.

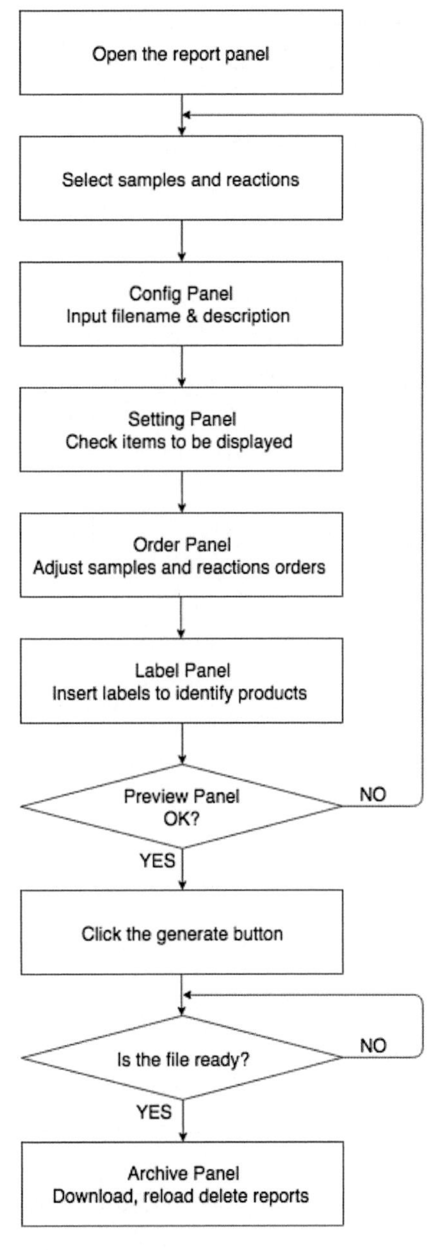

Figure 6-12. The flow chart for creating a report.

In the beginning, the report panel is displayed by clicking the green report button on the top navigation bar. There are three columns in Figure 6-11, and they are the collection column, the list column, and the detail panel from left to right. The report panel sits in the right column. Samples or reactions from different collections, which need to be picked in the middle column, are able to be added into one report.

In the report panel, there are six windows: “Config”, “Setting”, “Order”, “Label”, “Preview” and “Archive”. In the “Config” window, the report type, name, and description are inputted. Displayed items of samples and reactions are determined in the “Setting” window, in which the default setting shows all the information of subjects. Subjects order can be adjusted in the “Order” table by drag-and-drop targets. In the “Preview” window, the author can understand what is going to be shown in the report before generating it.

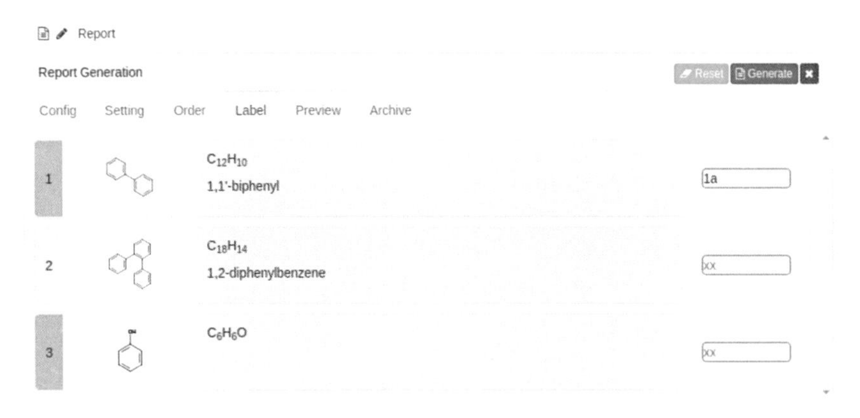

Figure 6-13. The “Label” window.

The manuscripts, supporting information and other reports in chemistry contain given structures along with labels of molecules to identify them. The reporting function meets this procedure by offering options for the assignment of a distinct label to each relevant molecule. This style improves the readability of given reports as the same molecule could be used in several reactions in different roles. This label automatically generates a clear structure of molecules as one in a report has only one label independent of how often the molecule appears in the report. The author can edit labels in the “Label” window, and a default value for a label is “xx”. These placeholders can be found and replaced after the report is generated. Figure 6-13 shows the interface to manage labels.

The “Archive” window, which is shown in Figure 6-14, lists all reports with names and descriptions. Tags containing types of documents append after file names. There are three buttons for each document: the orange editing button, the blue download button, and the red delete button. If the author wants to use the same setting as a previous report, the data can be retrieved by clicking the editing button without selecting subjects again. The author can continue to modify and generate a new report based on it.

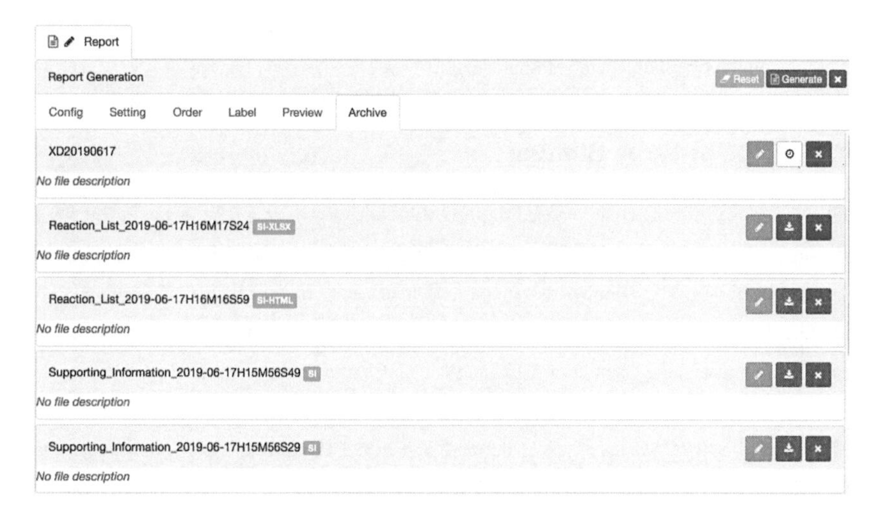

Figure 6-14. The “Archive” window.

Reports are generated when the system is in the low-loading status without impacting the server performance. When it is not ready, the download button is disabled and displayed in grey. A generated report is available to be downloaded when the download button is enabled in blue. By clicking the delete button, the report is removed from both the archive and the database.

6.3 Libraries

Different libraries or gems are developed and installed to realize the report module. The HTML template is enabled by the ERB class[110] in the standard ruby library, in which HTML elements and attributes are composed based on predefined layouts. Moreover, CSV files are generated from the ruby CSV class,[111] while the AXLSX gem[112] generates XLSX files. For Docx files, the Sablon gem[113] is chosen for a dynamic content generation from templates. These templates are Docx files that developers have to design in advance. Mail merge fields inside templates are replaced by user-defined data during the report generation.

Two crucial parts were not available when the report module was developed: the RInChI Ruby binding and the ChemDraw embedding.

6.3.1 RInChI Ruby Binding

The RInChI software is created based on the InChI software[114], which generates InChI for a single molecule, since a reaction combines several molecules. It is available in C++, but there is no Ruby library. The Simplified Wrapper and Interface Generator (SWIG)[115] is adopted to connect between Ruby and C++. Therefore, the rinchi-gem is developed to bridge the conversion between the Chemotion-ELN and the RInChI software.

In the rinchi-gem, a *MolVect* type is defined, and it is a string vector containing molecule molfiles. In the ruby code, three *MolVect* objects are created, which different roles of molfiles are inserted into starting materials, products, and reactants, respectively. Solvents are assigned to reactants. By inputting three objects of different role groups to the RInChI program, it returns a rinchi-string, Long-RInChIKey, Short-RInChIKey, and Web-RInChIKey to the ruby application.

Three RInChIKeys have different features. There is no fixed length for the Long-RInChIKey, which contains the most information of the reaction compared to the other two. The Web-RInChIKey is designed to fast search because of the shortest content, in the sacrifice of discarding more information. Table 6-2 compares the three RInChIKeys.

Table 6-2. Comparison of the three RInChIKeys.

RInChIKey	Length	Features
Long-RInChIKey	No fixed length.	- Recognizing molecules in a reaction with the standard InChIKey is possible. - uniqueness checks & exact search
Short-RInChIKey	63 characters	- exact searches
Web-RInChIKey	47 characters	- search

6.3.2 ChemDraw Embedding

ChemDraw[80] is a widely-used, commercially available cheminformatics tool. When a molecule is created in ChemDraw, it can be directly manually pasted to a Docx file rendering the molecule image. Moreover, by clicking this image in the Docx file, it triggers the ChemDraw software and restores the original structure. By editing in the ChemDraw application, the change is reflected in the Docx document.

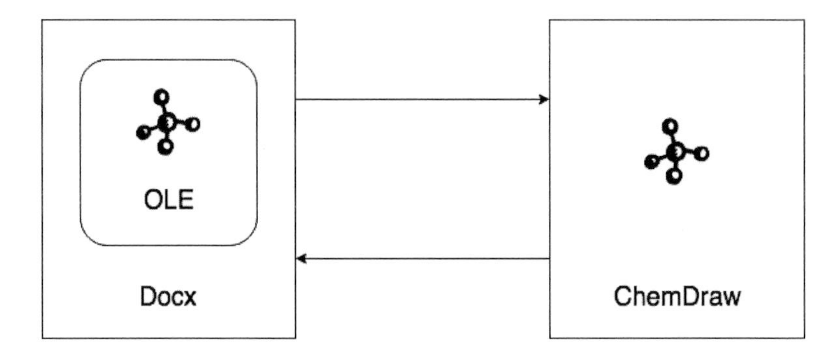

Figure 6-15. The bidirectional data flow between ChemDraw and a Docx document.

This bidirectional data flow between application and Docx document is an important feature, and it is accomplished by the OLE (Object Linking & Embedding);[116] which is developed by

Microsoft. Therefore, this feature is not available on Linux and Mac OSX systems. Files opened by a software other than Microsoft Office, e.g., Libre Office, have no access to OLE and ChemDraw. Figure 6-15 illustrates the bidirectional data flow.

The OLE object is built based on the ChemDraw CDX binary. Therefore, the ELN must have the ability to generate the CDX binary from given molecular molfiles. However, as a library meeting this purpose was not available, a conversion library had to be created and integrated into the ELN. This library can transform molfiles to the CDX format, and then, append binary digits to the OLE object.

ChemDraw has two formats to describe chemical structures: CDX and CDXML.[117] The CDX format consists of only binary data, and it is difficult for a human to read it. The CDXML format is similar to XML, which contains text information in the tree structure. Table 6-3 and

Table 6-4 list CDX and CDXML formats for the example alanine.

Table 6-3. Alanine in CDX format.

Format	Alanine
CDX only first 10 lines are displayed	566a 4344 3031 3030 0403 0201 0000 0000 0000 0000 0000 0080 0000 0000 0300 1300 0000 4368 656d 4472 6177 2031 362e 302e 312e 3408 0013 0000 0070 6865 6e79 6c61 6c61 6e69 6e65 2e63 6478 0402 1000 89de 3300 969f 9100 7621 9200 6960 3801 0109 0800 0000 0000 0000 0000 0209 0800 0000 0203 0000 0b02 0d08 0100 0108 0701 0001 3a04 0100 013b 0401 0000 4504 0100 013c 0401 0000 4a04 0100 000c 0601 0001 0f06

Table 6-4. Alanine in CDXML format.

Format	Alanine
CDXML	<?xml version="1.0"?> <**CDXML**> <**page**> <**fragment**> <**n** id="2" p="283.93 358.10"/> <**n** id="4" p="257.95 403.10" Element="7"> <**t** p="254.35 407"><**s**>NH2</**s**></**t**> </**n**> <**n** id="6" p="257.95 373.10" Geometry="Tetrahedral" /> <**n** id="8" p="231.97 358.10" /> <**n** id="10" p="283.93 328.10" Element="8"> <**t** p="280.03 332"><**s**>O</**s**></**t**> </**n**> <**n** id="12" p="309.91 373.10" Element="8"> <**t** p="306.01 377"><**s**>OH</**s**></**t**> </**n**> <**b** id="14" B="6" E="4" Display="WedgeBegin" /> <**b** id="15" B="2" E="6" /> <**b** id="16" B="6" E="8" /> <**b** id="17" B="2" E="10" Order="2" /> <**b** id="18" B="2" E="12" /> </**fragment**> </**page**> </**CDXML**>

The CDXML format is comprehensible. Atoms are defined in "n" tags, while "b" tags specify bonds. For example, in the alanine CDXML text, a wedge bond connects a carbon and an amine group: the bond id 14 is the wedge bond, which is defined in the "Display" attribute. It links atoms id 4 and id 6. The atom, id 4, is the nitrogen atom with atomic number 7 indicated by the "Element" attribute. Additional symbols displayed in the ChemDraw interface are also included in the wrapped "t" tag with positions, and "NH2" is shown in this case.

The corresponding CDX binary cannot be deciphered without predefined rules. In Table 6-3, the first 56 highlighted digits represent the XML header and the CDXML tag based on the ChemDraw transformation table.[117]

A process flow from reading molfiles to create the ChemDraw binding and to insert it in the Docx file is illustrated in Figure 6-16. Open Babel[71] can both read and write the CDXML text, but it can only read, not write, the CDX binary. ChemDraw provides rules of conversion between CDX and CDXML, which have to be inserted to the OLE file in the Chemotion-ELN. Therefore, the solution is to transform molfiles to the CDXML tree structure by Open Babel, and then, the CDX binary can be encoded from ChemDraw provided rules. In the end, the generated CDX binary is inserted into an OLE file, which can be embedded into the Docx file, by the ruby-ole gem.[118]

The transformation from CDXML to CDX is achieved by the *CDX::creator* class in the Chemotion-ELN. The Nokogiri gem[119] parses the CDXML tree structure, and there are six types of nodes: page, fragment, atom, bond, text, and arrow. The arrow node, not used within the sample transformation, is adopted in the reaction formation. Each node, identified by the node name, is converted to the corresponding string by merging default format and dynamic content, like position, element, charge, and isotope.

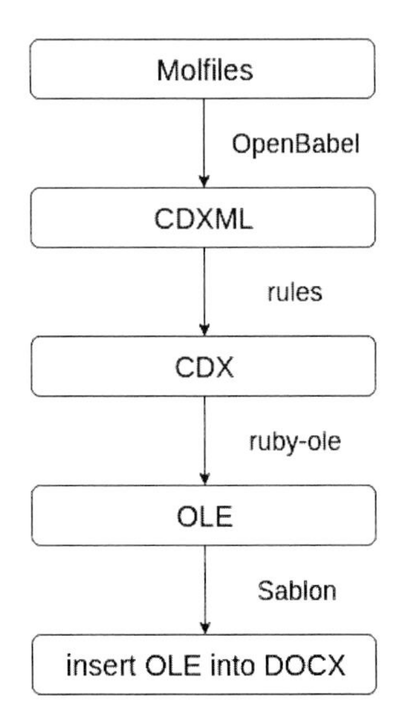

Figure 6-16. The process flow of creating OLE from Molfiles and inserting into Docx files.

Table 6-5. An atom node

CDX binary	**CDXML node**	**Function**
04 80	<n	Head of Node
11 00 00 00	id="17"	Node id
00 02 08 00 00 00 7B 00 00 00 46 00	p="70 123"	Node position
02 04 02 00 07 00	Element="7"	Element type
00 00	>	Tail of Node

The CDX format is hexadecimal with little-endian ordering, where the least-significant value is placed in the lowest-enumerated position. An atom node example is listed in Table 6-5. The tag name "n" is an atom node, which binary is 04 80 as defined in conversion rules. The id is an eight-digit number, and an example of it, id 17 is 11 in the first two digits. As to the node position, the first eight digits are fixed, 00 02 08 00, followed by two eight digits for y and x position. 02 04 02 00 represents the atom element, and four digits after it denotes the atom number. Therefore, a nitrogen atom is 07 00, since the atom number is 7. In the end, an ending, 00 00, is appended. Figure 6-17 illustrates a flow chart from the CDXML to CDX.

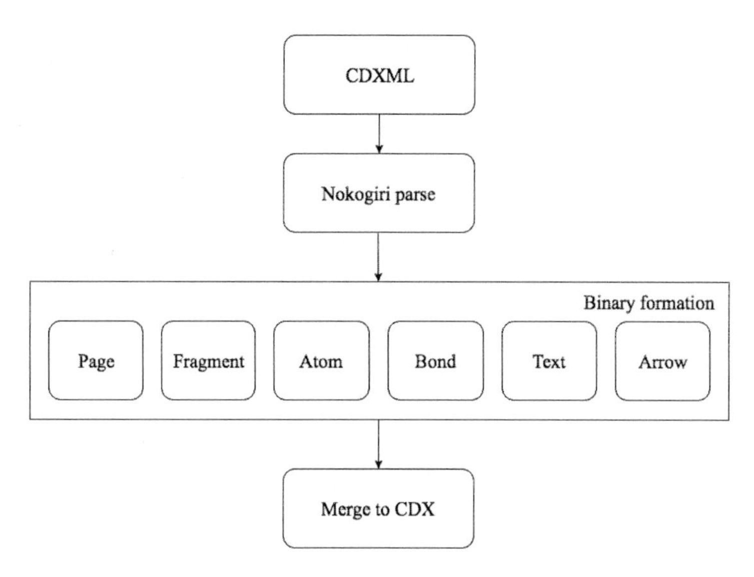

Figure 6-17. The CDXML to CDX conversion flow.

The Sablon gem is extended to add both image and OLE files which enables interaction among the Docx document and the ChemDraw application. Images are created by Inkscape,[120] a vector graphic software, from molecule structure SVG files. Since this process consumes higher computation resources, the delayed job gem[121] is integrated for asynchronous executions. Reports are prepared during low system loading.

6.4 Discussion

In this chapter, the report module built on the top of Chemotion-ELN, with the convenient working flow and essential functions for chemists, is presented. To sum up, two significant features that have not yet been implemented on other state-of-the-art ELNs are presented. First, generated reports meeting the FAIR data management standards. Second, ChemDraw binding brings in an essential operation which makes the created report preserve the advantage as done manually.

Following features in reports enable the FAIR standards:

(1) Findable: Every sample or reaction in reports is given unique and persistent identifiers, i.e. InChIKeys and RInChIKeys. Moreover, XLSX, CSV, and HTML formats, which will be added to publishers' servers and databases, facilitate searching on the internet.

(2) Accessible: Standard report and supporting information are defined in a human-readable format. The machine can read embedded ChemDraw objects. Meanwhile, XLSX, CSV, and HTML formats are structured data which is machine-readable.

(3) Interoperable: Content and format of reports, including spectroscopic data description, are verified by chemists in our group. Cheminformatics data in them is from the Chemotion-ELN, which integrates several standard tools and services, e.g., Open Babel and PubChem.

(4) Reusable: Publishers who received reports define the license. If owners release reports to the public, they can append license inside reports.

In the future, feedback from users using this system would be continuously analyzed and integrated.

7 Summary

In this thesis, three modules, namely (1) ChemSpectra, (2) reaction template and temperature predictions, and (3) reporting are developed, explored, and discussed. The work is a contribution to improve, facilitate, and accelerate scientific work in chemistry.

7.1 ChemSpectra

Up to date, the visualization and editing of essential analyses, i.e., IR, NMR, and MS, via web-viewers were limited to few functions. This deficiency confined the potential of chemical web-based infrastructure since an advanced spectra viewer and editor, embedded in a modern research data environment, is a prerequisite for efficient work with chemical data. The software ChemSpectra not only enables a complete platform for structure elucidation but also provides a solution to the state-of-the-art prediction of functional group existence.

The first function of the ChemSpectra, the spectrum editor, allows to visualize and edit IR, NMR and MS spectra in a modern web interface. Depending on the analysis type given and visualized, functions like threshold setting, solvent definition, peak picking, and zoom functions are enabled. The functions are not comprehensive with respect to a detailed analysis of scientific data but allow a first evaluation of the given spectra.

The second function of the ChemSpectra covers computer-assisted and deep learning methods to analyze the given data. Such a combination of spectra viewer/editor and automatic evaluation is currently not available. The computer-assisted and deep learning methods are realized in different ways. For IR, deep learning models are trained to infer the existence of functional groups from the spectrum profile. Currently, 36 out of the 44 most common functional groups are trained and deployed with balanced accuracies equal to or higher than 80%. In the future, by accumulating datasets from the Chemotion-ELN, a large number of functional groups will be included to increase the accuracy. For NMR data, the NMRShiftDB is connected to evaluate features from both ^{13}C and ^{1}H spectra semi-automatically. The future target, an end-to-end process, which has the potential to reduce error, is desired to decrease manual work effort. Automatic MS spectra selection from multiple scans is executed by identifying corresponding mass-to-charge ratio based on the estimated molecular structure. In the end, inference by combining the evaluation results of the four spectra types demonstrates a path to complete the quality assurance of chemical compounds.

7.2 Reaction Predictions

The improvement of deep learning methods for reaction prediction as well as retrosynthesis can increase research efficiency in synthetic chemistry. In this work, reaction prediction was applied with a focus on the extraction of the correct reaction template and the prediction of the most suitable temperature for a desired reaction.

In this thesis, an additional verification method is proposed to form the full template extraction procedure. Traditionally, the template extraction includes atom mapping, core extraction, and neighbor extraction. This procedure is currently applied without a detailed verification of the obtained templates. By the proposed procedure via restoration, it is possible to evaluate the extraction and to understand the deficiency of tools.

By analyzing the temperature of templates, a correlation is observed. A multi-layer perceptron receiving starting materials can infer the required temperature for reactions. The accuracy is around 73% if the correct prediction is defined within +/-10 °C error to the truth.

The template class prediction is around 77%, which is similar to published data.[45] Moreover, the model is planned to be improved by adding the temperature as an important condition to the outcome of a reaction. This work is the building block for further development of reaction predictions.

7.3 Reporting

The reporting on scientific work is very time-consuming. In this thesis, a feasible way to fast and effortless reporting in synthetic chemistry is demonstrated and implemented to the Chemotion-ELN. There are two major classes of the reporting module developed which are called "Standard report" and "Supporting information". These modules facilitate knowledge exchange among organizations based on machine-readable and structured information.

Chemists can migrate to this workflow with a ChemDraw binding, which was only done manually before, seamlessly with simple control.

The standard report is already adopted as an internal weekly review process in our chemistry group at the Karlsruhe Institute of Technology, for more than one year. The flow is verified and fine-tuned thanks to the co-work from chemists. It can be further customized by the modification of templates in the future.

The supporting information module is designed for publications which requires a more rigorous standard. This format meets the requirements of scientific chemistry journals and their traditionally chosen formatting of analytical and experimental data. Additional files, in XLSX, CSV and HTML formats, are used as data exchange media between organizations.

8 Abbreviations

%	Percent
°C	Temperature in Celsius
^{13}C NMR	Carbon Nuclear Magnetic Resonance
^{1}H NMR	Proton Nuclear Magnetic Resonance
1D	One Dimensional
2D	Two Dimensional
3D	Three Dimensional
ACS	American Chemical Society
A.I.	Artificial Intelligence
API	Application Programming Interface
BE	Backend
CAS	Chemical Abstracts Service
CDX	ChemDraw Exchange
CDXML	CDX in XML format
CID	Compound Identifier
CNN	Convolutional Neural Network
CPU	Central Processing Unit
CSS	Cascading Style Sheets
CSV	Comma Separated Values
DEPT	Distortionless Enhancement by Polarization Transfer
DL	Deep Learning
DNN	Deep Neural Network
ECFP	Extended Connectivity Fingerprint

ELN	Electronic lab notebook
FAIR	Findability, Accessibility, Interoperability, and Reusability
FCL	Fully Connected Layer
FE	Frontend
GCN	Graph Convolutional Neural Network
HTML	Hypertext Markup Language
HTTP	Hypertext Transfer Protocol
InChI	IUPAC International Chemical Identifier
IUPAC	International Union of Atomic and Molecular Physical Data
IR	Infrared
JCAMP	Joint Committee on Atomic and Molecular Physical Data
JSON	JavaScript Object Notation
KIT	Karlsruhe Institute of Technology
LSTM	Long-Short-Term memory
ML	Machine Learning
MLP	Multi-Layer Perceptron
MS	Mass spectrometry
NMR	Nuclear Magnetic Resonance
NN	Neural Network
OCR	Optical Character Recognition
OLE	Object Linking and Embedding
OSR	Optical Structure Recognition
QC	Quality Control
PDF	Portable Document Format

RDT	Reaction Decoder Tool
ppm	Parts per million
Rf	Retention factor
RInChI	Reaction InChI
RNN	Recurrent Neural Network
SI	Supporting Information
SMARTS	SMILES Arbitrary Target Specification
SMILES	Simplified Molecular Input Line Entry Specification
SVG	Scalable Vector Graphics
SWIG	Simplified Wrapper and Interface Generator
UI	User Interface
URL	Uniform Resource Locator
USPTO	United States Patent and Trademark Office
UV	Ultraviolet
WWW	World Wide Web
XML	Extensible Markup Language

9 Literature

1 Lee, J., Bagheri, B. & Kao, H.-A. A cyber-physical systems architecture for industry 4.0-based manufacturing systems. *Manufacturing letters* **3**, 18-23 (2015).

2 Ustundag, A. & Cevikcan, E. *Industry 4.0: managing the digital transformation.* (Springer, 2017).

3 Gilchrist, A. *Industry 4.0: the industrial internet of things.* (Apress, 2016).

4 Parviainen, P., Tihinen, M., Kääriäinen, J. & Teppola, S. Tackling the digitalization challenge: How to benefit from digitalization in practice. *International journal of information systems and project management* **5**, 63-77 (2017).

5 Kanza, S. *et al.* Electronic lab notebooks: can they replace paper? *Journal of cheminformatics* **9**, 31 (2017).

6 Kwok, R. How to pick an electronic laboratory notebook. *Nature* **560**, 269 (2018).

7 Herres-Pawlis, S., Koepler, O. & Steinbeck, C. NFDI4Chem: Shaping a Digital and Cultural Change in Chemistry. *Angewandte Chemie International Edition* (2019).

8 Tremouilhac, P. *et al.* Chemotion ELN: an Open Source electronic lab notebook for chemists in academia. *Journal of cheminformatics* **9**, 54 (2017).

9 Kotov, S., Tremouilhac, P., Jung, N. & Bräse, S. Chemotion-ELN part 2: adaption of an embedded Ketcher editor to advanced research applications. *Journal of cheminformatics* **10**, 38 (2018).

10 Potthoff, J. *et al.* Procedures for systematic capture and management of analytical data in academia. *Analytica Chimica Acta: X* **1**, 100007 (2019).

11 Engel, T. & Gasteiger, J. *Chemoinformatics: basic concepts and methods.* (John Wiley & Sons, 2018).

12 Engel, T. & Gasteiger, J. *Applied chemoinformatics: achievements and future opportunities.* (John Wiley & Sons, 2018).

13 Jordan, M. I. & Mitchell, T. M. Machine learning: Trends, perspectives, and prospects. *Science* **349**, 255-260 (2015).

14 Ramakrishnan, R., Dral, P. O., Rupp, M. & von Lilienfeld, O. A. Big data meets quantum chemistry approximations: The Δ-machine learning approach. *Journal of chemical theory and computation* **11**, 2087-2096 (2015).

15 Witten, I. H., Frank, E., Hall, M. A. & Pal, C. J. *Data Mining: Practical machine learning tools and techniques.* (Morgan Kaufmann, 2016).

16 Field, L. D., Sternhell, S. & Kalman, J. R. *Organic structures from spectra.* (John Wiley & Sons, 2012).

17 Jonas, E. & Kuhn, S. Rapid prediction of NMR spectral properties with quantified uncertainty. *Journal of cheminformatics* **11**, 1-7 (2019).

18 Kuhn, S. & Jonas, E. Rapid prediction of NMR spectral properties with quantified uncertainty. (2019).

19 Ito, K., Obuchi, Y., Chikayama, E., Date, Y. & Kikuchi, J. Exploratory machine-learned theoretical chemical shifts can closely predict metabolic mixture signals. *Chemical science* **9**, 8213-8220 (2018).

20 Polussa, A. *et al.* in *AGU Fall Meeting Abstracts.*

21 Gastegger, M., Behler, J. & Marquetand, P. Machine learning molecular dynamics for the simulation of infrared spectra. *Chemical science* **8**, 6924-6935 (2017).

22 Dührkop, K., Shen, H., Meusel, M., Rousu, J. & Böcker, S. Searching molecular structure databases with tandem mass spectra using CSI: FingerID. *Proceedings of the National Academy of Sciences* **112**, 12580-12585 (2015).

23 Hufsky, F. & Böcker, S. Mining molecular structure databases: Identification of small molecules based on fragmentation mass spectrometry data. *Mass spectrometry reviews* **36**, 624-633 (2017).

24 Elucidation, M. S. Mnova 12.0. 4, Mestrelab Research. *Santiago de Compostela, Spain, www. mestrelab. com* (2019).

25 McLean, A., Romain, C., Bakewell, C., Harvey, M. & Rzepa, H. Demonstration of Professional Preview of FAIR (Findable, Accessable, Inter-operable and Re-usable) NMR Data files using Mnova and Mpublish. *Imperial College Data Repository* (2016).

26 *Bruker TOPSPIN*, http://www.bruker/-biospin.com.

27 Menges, F. Spectragryph-optical spectroscopy software. *Version* **1**, 2016-2017 (2017).

28 Lancashire, R. J. The JSpecView Project: an Open Source Java viewer and converter for JCAMP-DX, and XML spectral data files. *Chemistry Central Journal* **1**, 31 (2007).

29 Mohamed, A., Nguyen, C. H. & Mamitsuka, H. NMRPro: an integrated web component for interactive processing and visualization of NMR spectra. *Bioinformatics* **32**, 2067-2068 (2016).

30 Vosegaard, T. jsNMR: an embedded platform-independent NMR spectrum viewer. *Magnetic Resonance in Chemistry* **53**, 285-290 (2015).

31 Beisken, S., Conesa, P., Haug, K., Salek, R. M. & Steinbeck, C. SpeckTackle: JavaScript charts for spectroscopy. *Journal of cheminformatics* **7**, 17 (2015).

32 Cornell, A., Belford, R., Bauer, M., Rothenberger, O. & Bergwerf, H. in *ABSTRACTS OF PAPERS OF THE AMERICAN CHEMICAL SOCIETY.* (AMER CHEMICAL SOC 1155 16TH ST, NW, WASHINGTON, DC 20036 USA).

33 *13C NMR Chemical Shift Table*, https://www.chem.wisc.edu/deptfiles/OrgLab/handouts/13-C NMR Chemical Shift Table.pdf

34 Bremser, W. HOSE—a novel substructure code. *Analytica Chimica Acta* **103**, 355-365 (1978).

35 Kuhn, S. & Johnson, S. R. Stereo-Aware Extension of HOSE Codes. *ACS Omega* **4**, 7323-7329 (2019).

36 Steinbeck, C., Krause, S. & Kuhn, S. NMRShiftDB constructing a free chemical information system with open-source components. *Journal of chemical information and computer sciences* **43**, 1733-1739 (2003).

37 Weininger, D. SMILES, a chemical language and information system. 1. Introduction to methodology and encoding rules. *Journal of chemical information and computer sciences* **28**, 31-36 (1988).

38 Weininger, D., Weininger, A. & Weininger, J. L. SMILES. 2. Algorithm for generation of unique SMILES notation. *Journal of chemical information and computer sciences* **29**, 97-101 (1989).

39 Michael Stephen Chen, Sophia Chen, Yanbing Zhu. (ed Stanford CS 229) (2017).

40 Affolter, C. & Clerc, J. Prediction of infrared spectra from chemical structures of organic compounds using neural networks. *Chemometrics and intelligent laboratory systems* **21**, 151-157 (1993).

41 Coley, C. W., Barzilay, R., Jaakkola, T. S., Green, W. H. & Jensen, K. F. Prediction of organic reaction outcomes using machine learning. *ACS central science* **3**, 434-443 (2017).

42 Do, K., Tran, T. & Venkatesh, S. in *Proceedings of the 25th ACM SIGKDD International Conference on Knowledge Discovery & Data Mining.* 750-760 (ACM).

43 Engkvist, O. *et al.* Computational prediction of chemical reactions: current status and outlook. *Drug discovery today* **23**, 1203-1218 (2018).
44 Quell, J. D. *et al.* Automated pathway and reaction prediction facilitates in silico identification of unknown metabolites in human cohort studies. *Journal of Chromatography B* **1071**, 58-67 (2017).
45 Segler, M. H. & Waller, M. P. Neural-symbolic machine learning for retrosynthesis and reaction prediction. *Chemistry–A European Journal* **23**, 5966-5971 (2017).
46 Warr, W. A. A Short Review of Chemical Reaction Database Systems, Computer-Aided Synthesis Design, Reaction Prediction and Synthetic Feasibility. *Molecular informatics* **33**, 469-476 (2014).
47 Wei, J. N., Duvenaud, D. & Aspuru-Guzik, A. Neural networks for the prediction of organic chemistry reactions. *ACS central science* **2**, 725-732 (2016).
48 Xiao, C., Zhang, P., Chaovalitwongse, W. A., Hu, J. & Wang, F. in *Thirty-First AAAI Conference on Artificial Intelligence.*
49 Schwaller, P., Gaudin, T., Lanyi, D., Bekas, C. & Laino, T. "Found in Translation": predicting outcomes of complex organic chemistry reactions using neural sequence-to-sequence models. *Chemical science* **9**, 6091-6098 (2018).
50 Segler, M. H., Preuss, M. & Waller, M. P. Planning chemical syntheses with deep neural networks and symbolic AI. *Nature* **555**, 604 (2018).
51 Duan, H., Wang, L., Zhang, C. & Li, J. Retrosynthesis with Attention-Based NMT Model and Chemical Analysis of the" Wrong" Predictions. *arXiv preprint arXiv:1908.00727* (2019).
52 Liu, X., Li, P. & Song, S. Decomposing Retrosynthesis into Reactive Center Prediction and Molecule Generation. *bioRxiv*, 677849 (2019).
53 Coley, C. W., Rogers, L., Green, W. H. & Jensen, K. F. Computer-assisted retrosynthesis based on molecular similarity. *ACS central science* **3**, 1237-1245 (2017).
54 Liu, B. *et al.* Retrosynthetic reaction prediction using neural sequence-to-sequence models. *ACS central science* **3**, 1103-1113 (2017).
55 Baylon, J. L., Cilfone, N. A., Gulcher, J. R. & Chittenden, T. W. Enhancing retrosynthetic reaction prediction with deep learning using multiscale reaction classification. *Journal of chemical information and modeling* **59**, 673-688 (2019).
56 Law, J. *et al.* Route designer: a retrosynthetic analysis tool utilizing automated retrosynthetic rule generation. *Journal of chemical information and modeling* **49**, 593-602 (2009).
57 Gao, H. *et al.* Using machine learning to predict suitable conditions for organic reactions. *ACS central science* **4**, 1465-1476 (2018).
58 Muegge, I. & Mukherjee, P. An overview of molecular fingerprint similarity search in virtual screening. *Expert opinion on drug discovery* **11**, 137-148 (2016).
59 Myint, K.-Z., Wang, L., Tong, Q. & Xie, X.-Q. Molecular fingerprint-based artificial neural networks QSAR for ligand biological activity predictions. *Molecular pharmaceutics* **9**, 2912-2923 (2012).
60 Landrum, G. Rdkit documentation. *Release* **1**, 1-79 (2013).
61 Morgan, H. L. The generation of a unique machine description for chemical structures-a technique developed at chemical abstracts service. *Journal of Chemical Documentation* **5**, 107-113 (1965).
62 Goodfellow, I., Bengio, Y. & Courville, A. *Deep learning*. (MIT press, 2016).
63 Goodman, J. (ACS Publications, 2009).
64 Jin, W., Coley, C., Barzilay, R. & Jaakkola, T. in *Advances in Neural Information Processing Systems.* 2607-2616.

65 Lowe, D. Chemical reactions from US pat.(1976-Sep2016), 2017.
66 Lowe, D. M. *Extraction of chemical structures and reactions from the literature*, University of Cambridge, (2012).
67 Lowe, D. & Sayle, R. in *ABSTRACTS OF PAPERS OF THE AMERICAN CHEMICAL SOCIETY.* (AMER CHEMICAL SOC 1155 16TH ST, NW, WASHINGTON, DC 20036 USA).
68 Mikolov, T., Karafiát, M., Burget, L., Černocký, J. & Khudanpur, S. in *Eleventh annual conference of the international speech communication association.*
69 Luong, M.-T., Pham, H. & Manning, C. D. Effective approaches to attention-based neural machine translation. *arXiv preprint arXiv:1508.04025* (2015).
70 Bahdanau, D., Cho, K. & Bengio, Y. Neural machine translation by jointly learning to align and translate. *arXiv preprint arXiv:1409.0473* (2014).
71 O'Boyle, N. M. *et al.* Open Babel: An open chemical toolbox. *Journal of cheminformatics* **3**, 33 (2011).
72 West, D. B. *Introduction to graph theory.* Vol. 2 (Prentice hall Upper Saddle River, NJ, 1996).
73 Defferrard, M., Bresson, X. & Vandergheynst, P. in *Advances in neural information processing systems.* 3844-3852.
74 Duvenaud, D. K. *et al.* in *Advances in neural information processing systems.* 2224-2232.
75 Kearnes, S., McCloskey, K., Berndl, M., Pande, V. & Riley, P. Molecular graph convolutions: moving beyond fingerprints. *Journal of computer-aided molecular design* **30**, 595-608 (2016).
76 Kipf, T. N. & Welling, M. Semi-supervised classification with graph convolutional networks. *arXiv preprint arXiv:1609.02907* (2016).
77 Chen, C., Ye, W., Zuo, Y., Zheng, C. & Ong, S. P. Graph networks as a universal machine learning framework for molecules and crystals. *Chemistry of Materials* **31**, 3564-3572 (2019).
78 Wilkinson, M. D. *et al.* The FAIR Guiding Principles for scientific data management and stewardship. *Scientific data* **3** (2016).
79 Filippov, I. V. & Nicklaus, M. C. (ACS Publications, 2009).
80 PerkinElmer. *ChemDraw*, https://www.perkinelmer.com/category/chemdraw
81 *Javascript*, https://developer.mozilla.org/en-US/docs/Web/JavaScript
82 *React.js*, https://reactjs.org/
83 *Python*, https://www.python.org/
84 *Flask*, https://flask.palletsprojects.com/en/1.0.x/
85 *D3.js*, https://d3js.org/
86 *SVG*, https://developer.mozilla.org/en-US/docs/Web/SVG
87 *jcampconverter*, https://github.com/cheminfo-js/jcampconverter
88 *Reselect*, https://github.com/reduxjs/reselect
89 Helmus, J. J. & Jaroniec, C. P. Nmrglue: an open source Python package for the analysis of multidimensional NMR data. *Journal of biomolecular NMR* **55**, 355-367 (2013).
90 Baumbach, J. I., Davies, A. N., Lampen, P. & Schmidt, H. JCAMP-DX. A standard format for the exchange of ion mobility spectrometry data (IUPAC Recommendations 2001). *Pure and applied chemistry* **73**, 1765-1782 (2001).
91 *SciPy*, https://www.scipy.org/
92 Kessner, D., Chambers, M., Burke, R., Agus, D. & Mallick, P. ProteoWizard: open source software for rapid proteomics tools development. *Bioinformatics* **24**, 2534-2536 (2008).

93 Adusumilli, R. & Mallick, P. in *Proteomics* 339-368 (Springer, 2017).
94 Kösters, M. *et al.* pymzML v2. 0: introducing a highly compressed and seekable gzip format. *Bioinformatics* **34**, 2513-2514 (2018).
95 *Matplotlib*, https://matplotlib.org/
96 *chemspider*, http://www.chemspider.com/
97 Ertl, P. An algorithm to identify functional groups in organic molecules. *Journal of cheminformatics* **9**, 36 (2017).
98 *RDKit IFG*, https://github.com/rdkit/rdkit/tree/master/Contrib/IFG
99 Plehiers, P. P., Marin, G. B., Stevens, C. V. & Van Geem, K. M. Automated reaction database and reaction network analysis: extraction of reaction templates using cheminformatics. *Journal of cheminformatics* **10**, 11 (2018).
100 *ChemAxon*, https://chemaxon.com/
101 Rahman, S. A. *et al.* Reaction Decoder Tool (RDT): extracting features from chemical reactions. *Bioinformatics* **32**, 2065-2066 (2016).
102 *Indigo Toolkit*, https://lifescience.opensource.epam.com/indigo/
103 Fatemeh, H., Ahmad, K. & Mohammad, D. M. ICMAP: An interactive tool for concept map generation to facilitate learning process. *Procedia Computer Science* **3**, 524-529 (2011).
104 Gonzalez, G. A. P. *et al.* Comparative evaluation of atom mapping algorithms for balanced metabolic reactions: application to Recon 3D. *Journal of cheminformatics* **9**, 39 (2017).
105 *BwUniCluster*, https://www.scc.kit.edu/dienste/bwUniCluster.php
106 Schneider, N., Stiefl, N. & Landrum, G. A. What's what: The (nearly) definitive guide to reaction role assignment. *Journal of chemical information and modeling* **56**, 2336-2346 (2016).
107 Huang, Y.-C. *et al.* Addition of dithi (ol) anylium tetrafluoroborates to α, β-unsaturated ketones. *Beilstein journal of organic chemistry* **14**, 515-522 (2018).
108 Grethe, G., Goodman, J. M. & Allen, C. H. International chemical identifier for reactions (RInChI). *Journal of cheminformatics* **5**, 45 (2013).
109 *Beilstein Institute*, https://www.beilstein-institut.de/en/home
110 *Ruby erb*, https://ruby-doc.org/stdlib-2.6.3/libdoc/erb/rdoc/ERB.html
111 *Ruby csv*, https://ruby-doc.org/stdlib-2.6.1/libdoc/csv/rdoc/CSV.html
112 *axlsx*, https://github.com/randym/axlsx
113 *Sablon gem*, https://github.com/ComPlat/sablon
114 Heller, S. R., McNaught, A., Pletnev, I., Stein, S. & Tchekhovskoi, D. InChI, the IUPAC international chemical identifier. *Journal of cheminformatics* **7**, 23 (2015).
115 *SWIG*, http://www.swig.org/
116 *OLE*, https://en.wikipedia.org/wiki/Object_Linking_and_Embedding
117 CambridgeSoft. *CDX File Format*, https://www.cambridgesoft.com/services/documentation/sdk/chemdraw/cdx/General.htm
118 *Ruby OLE*, https://github.com/aquasync/ruby-ole
119 *Nokogiri gem*, https://nokogiri.org/
120 *Inkscape*, https://inkscape.org/
121 *Delayed_job gem*, https://github.com/collectiveidea/delayed_job

10 Appendix

10.1 Accuracies of Functional Group Predictions

Here, Table 10-1, Table 10-2, and Table 10-3 show accuracies of the 44 most frequent functional groups in our dataset. The model architecture is defined in Table 4-14. 1D-CNN model architecture. 36 out of 44 functional groups existence predictions can achieve the accuracy equal to or larger than 80%.

Table 10-1. Accuracies of functional groups, from 1st ~ 10th.

	Functional group SMARTS	**Positive Test set count**	**Balanced accuracy (BACC)**	**BACC ≥ 90%**	**BACC ≥ 80%**
1	C-,:O	73	94.55%	Y	Y
2	C-,:O-,:C(-,:C)=O	51	96.81%	Y	Y
3	c-,:O-,:C	47	92.40%	Y	Y
4	c-,:[Cl]	40	84.62%		Y
5	c-,:O	33	88.40%		Y
6	C-,:[Cl]	31	82.78%		Y
7	c:,-n:,-c	29	82.74%		Y
8	C-,:C(-,:C)=O	28	91.83%	Y	Y
9	c-,:[N&+](=O)-,:[O&-]	23	93.89%	Y	Y
10	C=C-,:C	23	86.36%		Y

Table 10-2. Accuracies of functional groups, from 11th ~ 28th.

	Functional group SMARTS	Positive Test set count	Balanced accuracy (BACC)	BACC ≥ 90%	BACC ≥ 80%
11	C-,:O-,:C	20	91.56%	Y	Y
12	C-,:[Br]	20	80.12%		Y
13	c-,:[Br]	19	68.63%		
14	c-,:C(-,:C)=O	19	83.23%		Y
15	c-,:N	19	92.30%	Y	Y
16	C-,:F	18	91.50%	Y	Y
17	c:,-o:,-c	17	84.50%		Y
18	c-,:C(=O)-,:O-,:C	17	91.61%	Y	Y
19	C-,:C(=O)-,:O	16	95.77%	Y	Y
20	c-,:F	15	76.87%		
21	c=O	15	89.92%		Y
22	C-,:N-,:C	12	85.37%		Y
23	C-,:N(-,:C)-,:C	11	86.79%		Y
24	C-,:N	11	84.74%		Y
25	c:,-n(-,:c)-,:C	11	75.83%		
26	c-,:C(=O)-,:O	10	92.96%	Y	Y
27	C-,:C#N	10	71.72%		
28	c-,:C=O	9	91.18%	Y	Y

Table 10-3. Accuracies of functional groups, from 29th ~ 44th.

	Functional group SMARTS	Positive Test set count	Balanced accuracy (BACC)	BACC ≥ 90%	BACC ≥ 80%
29	c:,-[n&H1]:,-c	8	85.56%		Y
30	C-,:C(=O)-,:O-,:C	8	99.02%	Y	Y
31	C-,:N-,:C(-,:C)=O	8	94.94%	Y	Y
32	c-,:N-,:C(-,:C)=O	8	86.08%		Y
33	C=C(-,:C)-,:C	8	88.19%		Y
34	C-,:C#C-,:C	8	81.76%		Y
35	C-,:C=C-,:C	8	85.97%		Y
36	c:,-s:,-c	8	69.88%		
37	C#C-,:C	8	97.78%	Y	Y
38	c-,:C#N	8	76.75%		
39	c-,:C(-,:c)=O	7	78.17%		
40	c-,:N(-,:C)-,:C	7	88.39%		Y
41	C-,:C=C(-,:C)-,:C	7	80.18%		Y
42	c-,:N-,:C	6	87.15%		Y
43	c:,-n:,-n(-,:c)-,:C	5	88.14%		Y
44	c-,:I	5	75.95%		

10.2 Curriculum Vitae

Yu-Chieh Huang

Born 18.04.1981 in Taipei, Taiwan

E-mail: jasonych99@gmail.com

Github: https://github.com/JasonYCHuang

Professional Experience

05/2016 - 10/2019	**Scientist** Institute of Toxicology and Genetics, Karlsruhe Institute of Technology, Germany
07/2015 - 04/2016	**Software developer (Remote)** Andromoney
01/2006 - 04/2013	**Senior Electrical Engineer / Deputy Section Manager** **Defense Industry Reserve Duty** Foxconn Technology Group, Taiwan
07/2005 – 09/2005	**Research Assistant** Academia Sinica, Taiwan

Education

10/2013 - 03/2016	**M.S in Optics and Photonics** Erasmus Mundus - Europhotonics Karlsruhe Institute of Technology, Germany Aix-Marseille University, France
09/2003 - 07/2005	**M.S. in Engineering and System Science** National Tsing Hua University (NTHU), Taiwan
09/1999 – 06/2003	**B.S. in Mechanical Engineering** National Cheng Kung University (NCKU), Taiwan

Honor and Distinction

Ministry of Education Erasmus Mundus Award 2013

Ministry of Education, Taiwan

Research Creativity Award 2003

National Science Council

Outstanding Student Award 2001-2002 school year

NCKU

Publications

1. Huang, Y.-C.; Nguyen, A.; Gräßle, S.; Vanderheiden, S.; Jung, N.; Bräse, S., Addition of dithi (ol) anylium tetrafluoroborates to α, β-unsaturated ketones. *Beilstein journal of organic chemistry* **2018,** *14* (1), 515-522.

2. Tremouilhac, P.; Nguyen, A.; Huang, Y.-C.; Kotov, S.; Lütjohann, D. S.; Hübsch, F.; Jung, N.; Bräse, S., Chemotion ELN: an Open Source electronic lab notebook for chemists in academia. *Journal of cheminformatics* **2017,** *9* (1), 54.

3. Wei, P.-K.; Huang, Y.-C.; Chieng, C.-C.; Tseng, F.-G.; Fann, W., Off-angle illumination induced surface plasmon coupling in subwavelength metallic slits. *Opt. Express* **2005,** *13* (26), 10784-10794.

10.3 Acknowledgements

I would like to thank all those who contributed to the success of this work.

First and foremost, I would like to thank Prof. Dr. Stefan Bräse for the opportunity to work on the exciting interdisciplinary task as well as the great trust during the entire Ph.D.

With Prof. Dr. Ralf H. Reussner, I thank you for the friendly acceptance as the Korreferent.

Special thanks go to Dr. Nicole Jung, who always provides keen insight and patient guidance.

Thanks also go to the entire ComPlat team, both the software developers and chemists.

I would like to thank Dr. Pierre Tremouilhac, Dr. An Nguyen, Pei-Chi Huang, Chia-Lin Lin, Serhii Kotov, and Jan Potthoff. With collaboration, the Chemotion-ELN is getting more and more mature. I learn a lot from your fruitful discussion and feedback.

I would like to thank Sylvia Vanderheiden-Schroen, Simone Gräßle, Julia Kuhn, Dr. Anke Deckers, Dr. Patrick Hodapp, Dr. Nicolai Wippert, Dr. Steven Susanto, Jerome Klein, Jérome Wagner, and Laura Holzhauer. You provide crucial input and kindly share of chemistry experience to help us to improve the Chemotion-ELN.

My sincere thanks go to my family. I would like to thank my parents, Li-Mei Lin and Wu-Hsiung Huang, who made it possible for me to pursue my goals and dreams through their trust and support. I would like to thank my wife Chia-Hui Hung and two sons, Hao-Hsuan Huang and Chung-Hua Huang for your love, trust, and support.